# Progress or Freedom

Jean-Hervé Lorenzi · Mickaël Berrebi

# Progress or Freedom

## Who Gets to Govern Society's Economic and Technological Future?

Jean-Hervé Lorenzi
Paris Dauphine University
Paris, France

Mickaël Berrebi
Paris, France

*Translated by*
Dina Leifer
London, UK

ISBN 978-3-030-19593-9          ISBN 978-3-030-19594-6     (eBook)
https://doi.org/10.1007/978-3-030-19594-6

Cover illustration: Lev Dolgachov/Alamy Stock Photo
Cover design by eStudio Calamar

This Palgrave Macmillan imprint is published by the registered company Springer Nature Switzerland AG
The registered company address is: Gewerbestrasse 11, 6330 Cham, Switzerland

# Preface

## Progress or Freedom

This book is a plea for progress. It takes us deep into the world of new technology, with its extraordinary perspectives and its risks.

Digital technology is a hot topic today, with good reason, but this issue concerns many other scientific fields too, including genetics, energy and nanotechnology. Our freedom may be in danger: the leaders of these large technology firms want to define the world we live in for decades to come.

The issue, then, is to prevent companies from imposing their choices on the world, to the detriment of public authorities in all areas of our community and private lives.

One initial question emerges among many others: should we dismantle Google and the other big tech companies?

Paris, France

Jean-Hervé Lorenzi
Mickaël Berrebi

# Acknowledgements

We would very much like to thank Isabelle Albaret, Antoine Lefébure, Maurice Ronai and Guy Turquet de Beauregard for their valuable help with ideas, comments and suggestions.

We would also very much like to thank Angélique Delvallée for her constant support.

And finally, we would like to thank Marius Amiel, Pierre Garin, Léa Konini and Julien Maire for their kindness and their final proofreading.

# Contents

# List of Graphs

# List of Tables

# 1

# Introduction: The New Human Condition

The world is perplexed: a little lost, even. It is waking up to the fact that our emergence from the crisis does not in any way imply a return to the extraordinary growth of the early 2000s. It is finally realising that the ageing population, the demographic time bomb, the slowdown in productivity gains, the explosion in inequalities and unregulated finance are all creating entirely new economic conditions and, in fact, a slowdown in the world economy. Accommodative monetary policies are coming to an end, interest rates are set to rise again and fiscal policies, with the possible exception of Trump-style, temporary measures, are limited by the weight of public debts. We have reached a point today where the rational world is retreating and extremism and populism are rising, where the technological dream appears to be the only dream of a better world. This is what this book will discuss: the risks our societies are taking, with their naïve and simplistic view of a technological Eden: an Eden where politicians make way for the new prophets of technology, who are designing our world to suit themselves.

© The Author(s) 2019
J.-H. Lorenzi and M. Berrebi, *Progress or Freedom*,
https://doi.org/10.1007/978-3-030-19594-6_1

## 1.1    The Eternal Prophecy of a Better World

The technological illusion has a prophet: Jeremy Rifkin. He is a spokesman for great entrepreneurs who, despite their current promises, believe they can shape the world based on their innovations. Rifkin is far from the only one, of course. But he remains the most iconic figure, because he lends an air of scientific and cultural credibility to his views.

Why pick on this unfortunate propagandist for a world which is finally rid of all the hindrances we have endured for millennia: work, ignorance, wars and widespread change, beginning with the climate? Very simply, because he epitomises, on his own, the naïve world view whose keyword is "progress": a world where a sated and appeased consumer defines the new human condition. Rifkin conflates, under the general term "progress", science's remarkable developments and their technological applications for the majority of the population. But what precisely do we mean by "technological"? It can be defined as the sum total of individual processes designed for production, and therefore as the result of a concrete application of science, science being our tool for understanding the world. All scientific processes indicate an expertise which claims to be perfect, rigorous, increasingly concerned with regulation, which bases that claim on a heightened use of previously unknown computational tools. Technologies, and subsequent innovations, are nothing more than applications of these great advances in knowledge. And it is from this confusion that the problem is born.

Let's go back to Rifkin. His work, *The Zero Marginal Cost Society* (Rifkin 2014), pulls off the coup of making the entire Internet the answer to the crisis in the capitalist system and the threats it poses to humans and the environment. How better to resolve mass unemployment, or even "the end of work", as Rifkin has long described it, than by imagining "prosumers", capable of producing everything they need? How better to do away with our obsession with the hypothetical notion of growth, and to resolve the now central problem of inequality, than by envisaging a peer-to-peer, sharing, collaborative society, where profit no longer has any meaning? A society which can spread through the poorest regions of the world, as is the case in certain rural communities in India? How better to recreate a common good than by imagining a new model of governance, "collaborative commons", with a nod to the "commons" of feudal times, where production for use predominates over production for exchange? Finally, entering the realm of false assumptions, how better to reduce humanity's carbon footprint that by promoting renewable energy and a lifestyle which reconciles "free everything" abundance with sustainability?

Rifkin contends that the world is heading towards a third industrial revolution, based on the Internet of Things. But can we safely state this is an industrial revolution, in the sense of a new balance between production and consumption, creating a new cycle of economic growth and resulting from a series of innovations related to the boom in, and distribution of, new technologies? The conclusion is risky, because the development of the Internet of Things and of renewables remains embryonic and uncertain.

But most importantly, there is no consensus on this misused concept of industrial revolution. Once again, it is Schumpeter who puts us back on the right track: "if we survey the course of economic history, we do not find any sudden ruptures, only a slow and continual evolution" (Schumpeter 1946). Economists and historians have always been in a constant dialogue over the dynamics of technological change. Some, like Braudel, see it as a linear process, whereas others favour the disruptive approach. The idea of the industrial revolution, which is the result of the second approach, must be handled with care.

The uncertainty around the theory of a third industrial revolution is not just technical, moreover. The development of an Internet of renewable energy presupposes a collaborative economic approach which supersedes the traditional mode of production based on market exchange. Whether it is a question of advances in technology, or in the mode of production and consumption that these technological developments presuppose, it is questionable whether the conditions for an industrial revolution have been met.

Despite such a debatable approach, Rifkin, the prophet, a kind of heir to Charles Fourier and his Phalansteries, is right on target in a world full of nightmare scenarios. His offering of such naïve optimism has seduced quite a few people.

But if he was the only one, the world would be simple and criticism easy. In fact, he is joined by other prophets: those who, not content with conference speakers' fees, share their vision of the world from their position at the heart of the current economic establishment. Listen to them: Eric Schmidt[1] (Seigler 2010) explains: "Your car should drive itself. It's amazing to me that we let humans drive cars…". Similarly, Jeff Bezos (Quinn 2015)[2] reckons that the task of delivering parcels will be done by drones, so that: "One day, (such)… deliveries will be as common as seeing a mail truck". And what about Sundar Pichai (Tung 2016),[3] the man said to play Moses to Larry Page[4]'s God, by deciphering abstract projects from a mind too brilliant to be understood by everyone? He says: "the very concept of the 'device' will fade away. Over time, the computer itself, in whatever form, will be an intelligent assistant helping you through your day". As for the fascinating Elon Musk (Musk 2017),[5] he is quite determined to create entirely self-sufficient cities on Mars, because: "if we stay on Earth forever, there will be some eventual extinction event".

These are exceptional men: remarkable innovators and industrialists. But for all that, should they be the ones pointing the way forward for humanity? A humanity fascinated by new tools, overcome with gratitude towards those who provide them for us; a humanity fascinated by extraordinary means of communication, yet distraught when faced with an unfathomable world? Deep down, it can all be summed up by the simple idea that progress is never-ending, that it applies to everyone everywhere, that it transforms and improves our lot and that it is appropriate that those who design it should also set the rules.

Thus, artificial intelligence and gene technology would be tools in the hands of all-powerful demiurges. Based on their current economic power, they would naturally qualify as the sole architects of a recreated world. This would spell the end of thinkers on the nature of human progress, such as John Rawls on fairness and Amartya Sen's capabilities approach, on development for all; an end to the women and men who could alert the world to climate risks; an end to the Mandelas and others who could pave the way to peace in a violent world. From now on, only Mark Zuckerberg,[6] Larry Page and others like Sergey Brin[7] will have a voice. But, as ever, how much of this is new?

Are the misuse of technology and the pronouncements of these prophets unique in human history?

## 1.2 The Recurring Conflict Between Progress and Society

This debate is not really new. In fact, the dominant schools of thought have been confronting each other for centuries: those who control disruptive technologies in order to shape tomorrow's society, and those who think the power to make society progress belongs to those who conceive it in human terms. We need only consider the strong reluctance of the great thinkers in relation to the concept of progress: Paul Valéry said: "Modern man is the slave of modernity; there is no progress which does not turn into his complete servitude" (Valéry 1948). Technology against humanities: it is an eternal conflict, because power's only real prize is to make the rules which govern the lives of those who follow us.

In the past, economists perceived technological progress as an exogenous variable and declared they were not competent to analyse it. In fact, Lionel Robbins wrote that "Economists are not interested in technique as such" (Robbins 1932). Even Pareto excludes technological development from economic logic and considers it as external, gratuitous data in his model.

But economists did not remain absent from this arena. Innovation gradually becomes one of the principal levers of growth, and the cycles of innovation and economic growth are brought closer together, in the manner of Kuznets, for whom: "several periods of economic growth in the modern age can be identified with major innovations and the relative growth of the industries concerned" (Kuznets 1973). It is well known that this development in economic thought finds its most complete expression in Schumpeter, for whom technological progress is the engine of history and innovation the engine of growth. In fact, the influence of technological progress on economic growth and development appears to be firmly established, although perhaps not entirely so. Let's remind ourselves of Jacques Ellul, the undisputed technological thinker par excellence, little known because he was undoubtedly ahead of his time. According to Ellul, in every aspect of technology, it is really a human drive which is at work: the drive of power. "Technology is power, made up of instruments of power, hence producing phenomena and structures of power, i.e. of domination" (Ellul 2018). This impulse has found very different applications throughout the ages, which we should bear in mind, because it gives us hope that the future has not yet been decided.

Let's go back ten centuries. In the year one thousand, Europe is lagging behind, widely outpaced by the Chinese and Islamic societies and civilisations. The former has already seen the emergence of gunpowder, the compass, paper pulp and printing. The latter produced algebra and new advancements in medicine. But those technological innovations, produced by an educated elite, remain within the social circles of the powerful dynasties as they rise and fall. Take for example the clock invented by the Buddhist Monk and mathematician Yi Xing. It was exhibited at the emperor's palace, no less, in 725, but was eventually sidelined for lack of maintenance. Europe begins its "first industrialisation" as Jean Gimpel rightly says, in the eleventh century, with the spread of the new energy source: windmills, along with seed selection and the forge. But according to the historian Georges Duby, the advance may be due to Christianism, which is a religion of history, keen on progress. Paradoxically, the same is true of weaknesses in centralised power, whether religious or secular. Christian schisms, such as that of Saint Bernard in the twelfth century, spread technical skills through the rural world. Closer to our times, the thinkers of the Enlightenment would be right about absolutism and open the way for the industrial revolution, which begins in eighteenth-century England. If we learn anything from this brief recollection of history, it is that power relationships around technology have not always been the same and that technology was often seized by the majority against the wishes of an authority or a system of power.

Certainly, people who think of the future think of progress. But as Ellul reminds us, technology is not good or evil, but ambivalent. Saint-Simon, who only sees human development through the development of industry, is answered by Jules Vallès, who in 1848 declares himself to be the representative of poverty and of those without status, the proletariat. Science and technology: do they mean the liberation or the enslavement of mankind? It is an eternal debate and an eternal conflict between those who believe in fairness and those who believe in utility, not forgetting the iconoclasts, who do not accept this dualism. Let us think of the German Herbert Marcuse who writes: "The liberating power of technology – the manipulation of things – becomes a barrier to liberation and turns to the manipulation of people" (Habermas 1978). Films such as Fritz Lang's *Metropolis* (1927) and Charlie Chaplin's *Modern Times* (1936) illustrate the importance of technological progress and at the same time the enslavement of the masses it produces. People are naturally wary and if we believe Ellul, they are not wrong to be so. Technology does not consist of a simple accumulation of machines, but in the legitimate search for the most efficient means of production in all sectors. It is therefore put to use as much in the material as in the virtual world and ultimately structures the way we live as a society. Anthropologists refer to this technical age we live in as one which may have hindered mankind's freedom of action and judgement. It is a bleak assessment which they put down to the liberation of technology, which has become independent of social organisation. Or to put it another way, it has become independent of the economy, politics, culture, morality—in short, of humanity. This reading recalls the works of Andre Leroi-Gourhan, not in his conclusions, but in his forecasts: "This relationship between manual technicality and language[…] is certainly one of the most satisfying aspects of palaeontology and psychology, because it re-establishes deep links between gesture and word, between thoughts which can be expressed and the creative activity of the hands" (Leroi-Gourhan 1983).

We are therefore witnesses to a permanent conflict between progress and society. Who will win it in the coming years?

## 1.3 Who Will Shape the Twenty-First Century?

All of this appears very far off. We dream of escaping the powerful domination of material things over our minds. We are convinced that scientific and technical progress has been tamed, once and for all today; that its early twenty-first century architects are simply advocates of a peaceful revolution. This is nothing but pure naivety, for never in human history has the eternal challenge of our condition, the power some exercise over others, been so strongly concentrated in the hands of the creators and experts. This leaves the gains of the last few centuries, of free thinking and democracy, in tatters. Just look at the debate on climate and the major risk to humanity from deliberate extinction. Without going as far as the slightly extreme views of Ulrich Beck, for whom global society is a "risky manufacturer" (Beck 1992), whose troubles are deep-rooted and whose dangers have no geographical, temporal or social limit, we can subscribe to his statement as a concise indictment: "the system of regulation which is supposed to ensure the 'rational' control of these current potential causes of self-destruction is as useful as a bicycle brake on a jumbo jet" (Beck 1992).

Without ever losing sight of the success of the last few decades, when a middle class emerged that left poverty behind, we need to know where tomorrow's power lies. We have very legitimate reasons to fear. One of the most iconic scientists, Stephen Hawking, believed: "the development of full artificial intelligence could spell the end of the human race" (BBC 2014). That is why researchers like Laurent Orseau and Stuart Armstrong are working on the development of a "red button", a system aimed at preventing artificial intelligence from defying Isaac Asimov's second law of robotics, avoiding any act of rebellion by a machine if it decides to stop obeying humans. Scientists have also voiced anxieties over the question of the human genome. When the team led by Junjiu Huang (Cyranoski and Reardon 2015)[8] attempted to modify the genome of a human embryo in 2015 using a new technique[9] to prevent the development of a disease, the experiment also carried the risk of changing human heredity, no longer just one part of the faulty cells. Many scientists mobilised to highlight the ethical and social implications of this ill-considered technological advance, including 2015 Nobel Prize winners for medicine David Baltimore and Paul Berg. Previously, correcting the genome remained highly complicated, but today this no longer seems to be the case.

So we understand where the problem lies. Of course, we must free ourselves from onerous work; of course, we must find genetic solutions for previously incurable diseases and deformities, and there is no doubt these constitute major developments in human history. But that is not the problem. The problem is to determine who will set the limits on artificial intelligence, on genetic transformation, on the use of private data and so on.

The problem has never been so concrete, so fundamental, before. Its violence is more intellectual than physical.

## 1.4    The Fear and Hope of an Expectant World

We will try to describe this entire conflict; we will raise its risks and inspire its hopes. Obviously, we will not limit ourselves to describing the forerunners of a new scientific and technological revolution. The 3D printers, smartphones and so on are just a primitive representation of a world described as disrupted. In reality, the fundamental upheavals are still to come. The man nicknamed "the modern Thomas Edison", Raymond Kurzweil, a highly influential futurologist from MIT and a Google employee, is undoubtedly one of the most prolific forward thinkers. His list of predictions is long: it extends all the way to 2099. He describes the different stages which will lead humans towards a new kind: the "augmented" human, or half human, half robot. He is an enthusiastic supporter of Moore's law and estimates that computers will reach human-level intelligence by 2029. But behind all of that, his first and foremost objective is to postpone the age of death, with the eventual aim of making humans immortal. But that is all very far off.

Today, the main risk is that employment will become truly polarised. We may see high-skill jobs involving 1–10% of the population alongside "bullshit jobs" and a relative decline in the middle class, exactly as Daniel Cohen (2016) described: "At the very top, we find 'superjobs' for the top 1–10% of the population, who have grabbed half the economic growth for themselves alone. At the very bottom we find the 'bullshit jobs', the ones nobody wants, in construction, back kitchens and refuse. Only immigrants will accept these jobs, because it is their entry ticket to society. And in the middle, a working class which has undergone deindustrialisation, and a lower middle class which has lost all hope of advancement, because software has made all the intermediate jobs it filled redundant: jobs which used to form a link between the top and the bottom of society".

This totally unprecedented situation leads to the creation of what Pierre-Noël Giraud calls "useless men" (Giraud 2015). A new form of working class is trying to escape this label, seeking at any price to fit into the society which excludes them, as Joan Robinson stated, because: "The misery of being exploited by capitalists is nothing compared to the misery of not being exploited at all" (Robinson 1962). How distant from our dream of liberating Fordism…

And as ever, words are supreme. Words define good, evil, progress, advancement, improvement and the world to come. We are warned about impending automation and rightly so. In 2013, Carl Benedikt Frey and Michael Osborne (2013) announced that 47% of jobs in the United States were susceptible to being replaced by robots in the next ten or twenty years. Now, it is the turn of the OECD to produce a statistic of the same order of magnitude. According to the OECD, robots threaten to replace 40% of workers who are not educated to 'A' level or equivalent. We read about the incredible human creativity in the software sector and that is exciting. We are told about developments in medicine and that is hugely satisfying. We are delighted about the widespread lengthening of a healthy lifespan. But at the same time the world is becoming sterile, divided, fragmented, distanced from death and therefore from life by this stupid dream of an immortal human.

We hope our approach (neither optimistic nor pessimistic, only voluntaristic) in affirming the primacy of humans over machines and the consideration of rational arguments over prophecy is conducted in a rational and convincing manner. First of all, we must return to the argument over the development of the world economy confronted with this technological progress, and present it as objectively as possible. In *A Violent World* (Lorenzi and Berrebi 2016), we signalled the slowdown in the world economy. But it is not, as some people think, permanent. Next, we will try to show that technological disruption is only in its first stages and what is at stake in the coming years is far more important than providing a modern world framed only in terms of digital communication tools. We will try to rediscover the human being, with his or her insatiable need to feed, care, educate and shelter him or herself, and therefore to work. We often feel we are being sold a different human being: a superhuman in charge of permanently connected objects. We are sorry to say this new human is actually designed by our current masters of technology. Meanwhile, society is reforming itself, bringing inequalities the like of which have rarely been seen for two centuries, and in which the mastery of technology, carefully differentiated between some and others, imposes strict divisions on a society in social decline. So who will decide on the development of these societies? The tech giants, who know everything about our status and our lives today, via still-basic digital technology, through what can only be called widespread spying? Or the giants of human history, the great thinkers who have always managed to restore the humanity of societies which sometimes lose their way?

And that is the entire objective of this book: to offer an alternative to a world dominated by technology and its prophets: a world where technology is led by humans and by a definition of progress which holds fulfilment for all as the cardinal virtue of a progressive society.

# Notes

1. CEO of Google from 2001 to 2011. Siegler MG (2010) Techcrunch. Available via https://techcrunch.com/2010/09/28/schmidt-on-future/.
2. Founder of Amazon and of the aerospace company Blue Origin. Quin J (2015) Jeff Bezos: Five Things We Learned from the Amazon Founder. *The Telegraph*. Available via https://www.telegraph.co.uk/technology/amazon/11800416/amazon-founder-jeff-bezos-what-we-learned.html.
3. Appointed CEO of Google in 2015. Tung L (2016) ZDNET. Available via https://www.zdnet.com/article/google-ceo-pichai-says-devices-will-fade-away-but-launches-new-hardware-division/.
4. Co-founder of Google.
5. Founder of SpaceX (astronautics and space flight) and cofounder of PayPal. Musk (2017) Mary Ann Liebert, Inc. Available via https://www.liebertpub.com/doi/full/10.1089/space.2017.29009.emu.
6. Co-founder of Facebook.
7. Co-founder of Google with Larry Page.
8. Sun Yat-Sen University, Guangzhou, China. Cyranoski D & Reardon S (2015) Chinese Scientists Genetically Modify Human Embryos.

*Nature*. Available via https://www.nature.com/news/chinese-scientists-genetically-modify-human-embryos-1.17378.
9. CRISPR-Cas9.

# References

BBC. (2014). *Stephen Hawking Warns Artificial Intelligence Could End Mankind*. Available via https://www.bbc.co.uk/news/technology-30290540.

Beck, U. (1992). *Risk Society: Towards a New Modernity*. London: Sage.

Frey, C. B., & Osborne, M. A. (2013, September). *How Susceptible Are Jobs to Computerisation?* Oxford Martin Programme on Technology and Employment. https://www.oxfordmartin.ox.ac.uk/downloads/academic/The_Future_of_Employment.pdf.

Cohen, D. (2016, November 19). L'élection de Trump? Une infinite demande de protection? *Nouvel Observateur*. Paris.

Ellul, J. (2018). *The Technological System* (J. Neugroschel, Trans.). Oregon: Wipf and Stock.

Giraud, P. N. (2015). *L'Homme inutile – Du bon usage de l'économie*. Paris: Odile Jacob.

Habermas, J. (1978). *La Technique et la science comme "idéologie"*. Paris: Gallimard coll. Tel.

Kuznets, S. (1973). Innovation and Adjustments in Economic Growth. *The American Economic Review, 63,* 247–258.

Lorenzi, J. H., & Berrebi, M. (2016). *A Violent World*. London: Palgrave.

Leroi-Gourhan, A. (1983). *Mecanique vivante*. Paris: Fayard.

Robinson, J. (1962). *Economic Philosophy*. Oxford and New York: Routledge (2017).

Rifkin, J. (2014). *The Zero Marginal Cost Society*. London: Palgrave Macmillan.

Robbins, L. (1932). *An Essay on the Nature and Significance of Economic Science*. London: Macmillan. https://mises-media.s3.amazonaws.com/Essay%20on%20the%20Nature%20and%20Significance%20of%20Economic%20Science_2.pdf?file=1&type=document.

Schumpeter, J. A. (1946). Capitalism. In R. V. Clemence (Ed.), (2003). *Essays on Entrepreneurs, Innovations, Business Cycles and the Evolution of Capitalism* (pp. 189–210). Piscataway, NJ: Transaction Publishers.

Valéry, P. (1948). *Reflections on the World Today* (F. Scarfe, Trans.). Michigan: Pantheon.

# 2

# A Major Stagnation, But Not a Secular One

Why get involved in this ongoing debate among mostly American economists, just when we want to picture ourselves in science fiction-like visions of the world beyond tomorrow? Because it is precisely this high-level confrontation that brings us back to the real world of today and tomorrow.

What a strange expression "secular stagnation" is! Firstly because how can we seriously imagine today what will happen at the end of the century? Obviously we cannot, but the expression is a reaction against the naïve outlook of our Western societies, which refuse to contemplate any other scenario than the politically correct one, termed "progress". This outlook excludes any vision of the world except one which imagines ever-growing improvement in living standards for all, broadening of knowledge and democratisation of all current political regimes. The question of the survival of humanity and climate change is out of the picture. The pseudo-wars between civilisations are away in the distance; people lacking water and electricity are forgotten. Terrifying conflicts and a Mediterranean transformed into an immense graveyard are blanked out. As ever, the truth actually lies between the two visions: one naïve, the other deathly. And the object of the following pages is to

© The Author(s) 2019
J.-H. Lorenzi and M. Berrebi, *Progress or Freedom*,
https://doi.org/10.1007/978-3-030-19594-6_2

establish how politicians must strive for a realistic improvement in our living standards. This vast issue is affected negatively and positively by digital, environmental and demographic changes, and by breakthroughs in the fields of energy, genetics, information technology and astrophysics. These factors also lead us to reflect on possible new forms of growth.

## 2.1    Secular Stagnation, or the Ways of Freedom?

It is the key debate among economists today. It would have been unimaginable a few years ago, in the early euphoria of the Internet, and is only emerging today because a large number of highly acclaimed economists from across the Atlantic have pitched in with completely contradictory positions. Is this really such a new thing? And if it is not, how do we build on the lessons of the past? Of course, there is the actual birth of the concept of stagnation. It is already present in Keynes' benevolent gaze on Malthus and his dark vision of human development. He questions the very pursuit of growth, that orthodoxy of modern times, in the fascinating way that he links it to demography. This is the very issue raised by the great, unprecedented crisis of 1929, that black hole in the history of capitalism between the Great Depression and the Second World War, which brought the new scourge of mass unemployment with the ancient scourge of poverty. What if Malthus was also right at that time about unemployment? It is an approach which Keynes explores in 1937 (Keynes 1937) and which his American disciple follows in 1939 (Hansen 1939). The former writes that a stable population can only help improve living conditions for all and recalls, with a touch of humour, that countries have a new spectre, at least as ruthless as Malthusianism: the demon of unemployment caused by the breakdown in demand.

It is in this era, in 1938, that Alvin Hansen first mentions "stagnation" in relation to the demographic deficit of the United States and calls on public authorities to support demand in order to avoid the worst outcome. This Keynes disciple is simply returning here to his British mentor's thinking. Keynes' concern at the new mass unemployment in 1937 amounted to a denunciation of poverty: "wretchedness", the scourge which has provoked humanity's deepest fears throughout the ages.

Alvin Hansen recalls that European population growth in the nineteenth century is unprecedented throughout history and continues that way until the First World War. The United States experiences the same phenomenon in the decade following that war. In his view, the slowdown he observes in 1939 is not altogether bad news and thus he rehabilitates Malthus, by mentioning the insoluble problems which could be posed by too strong a growth in population. Nevertheless, he says, what he calls this "radical" decline in demographic growth, must be accompanied by public policies to support demand, to avoid the United States entering a lasting period of economic stagnation.

Above all, let us not forget the famous debate between Paul Sweezy (1942) and Joseph Schumpeter (1942). These two friends savaged each other via interposing books and articles. Sweezy, naturally, is convinced of a deep stagnation, unlike Schumpeter, who is convinced of the cyclical nature of the economy, linked to great waves of innovation. That sums up their positions and how they express them. But if we return to their current positions today, we see that in reality, the two men centre their arguments respectively on supply and on demand and this still explains their differences.

First, there is supply, with the likes of Tyler Cowen and Robert Gordon. Then demand, with Larry Summers, Paul Krugman, and many others. Then the overview orchestrated today by economists like Barry Eichengreen and Edmund Phelps. But the real question remains. Why has this debate acquired so much importance and legitimacy? Quite simply because they have all finally admitted that the widespread decline in productivity gains for the past several years all over the world has no other realistic explanation than the current patterns of technological progress; because the incredible polarisation of the labour market between high-skill, well-paid jobs and a kind of new proletariat calls into question the idea of improved knowledge for everyone; and because the slowdown in growth that we flagged up three years ago (Lorenzi and Berrebi 2016) has finally brought the broad mass of econoniists towards a greater realism and a better understanding of the current reality. But as always, we come back to secular (i.e. long term) visions, as though the word "transition" did not exist. Let us return to the arguments on both sides.

Firstly, what unites the different approaches to secular stagnation is that they all agree that it describes a situation characterised by weak inflation and near-zero interest rates. But beyond that common definition, the approaches clash. As Barry Eichengreen states (Eichengreen 2014): "Secular stagnation, we have learned, is an economist's Rorschach Test. It means different things to different people". Each one may well have expounded their theory on secular stagnation, but that does not mean they all understand the concept in the same way. The supply side explains secular stagnation via the slowdown in potential growth. In brief, growth is weak because potential growth[1] itself has slowed down! For Robert Gordon, it is not a case of saying that technological progress has stalled, but rather that the growth in technical progress is going to return to its historical level, which is very low. Above and beyond technical progress, he says there are six structural headwinds, including: the ageing population, mass training reaching a level where we can expect no more from it, rising inequality denying all income development for the middle classes since the 1980s and a level of debt which has become unsustainable (Gordon 2012). These constraints would explain the slowdown in productivity and in potential growth. He believes the great inventions of the past are of a wholly different nature to those we see today. Electricity, the internal combustion engine, running water, etc. are innovations which are far more important and productive than those brought by new technologies linked to the Internet, which may have more of an impact on our behaviour as consumers than on productivity. In fact, not only have developed countries failed to continue to invest in infrastructure, education and training, but they must also face up to demographic ageing. In Europe, for example, the dependence ratio, that is the number of retired people in relation to people of working age, is set to increase from 20.3% in 2000 to 35.4% in 2025. And this trend will most definitely foster the slowdown in labour input. This debate, which is so fascinating, will develop on that basis.

For example, Joel Mokyr considers that it is difficult to measure the real contribution of technology to productivity since indicators like GDP and factor productivity "were designed for a steel-and-wheat economy" and are therefore no longer at all suitable for our new economy. Another line of argument, by Barry Eichengreen, states it is all a question of time in the end. The weakness in productivity is transitory and productivity gains will materialise as soon as the system adapts and is prepared to fully exploit the potential of new technologies. For Erik Brynjolfsson and Andrew McAfee, real and extraordinary disruptions are to be expected from technologies like artificial intelligence, Big Data, robotics, driverless cars and also from the field of medicine. Are they right or are they wrong? Only one thing is certain: no technical system can be disassociated from the conditions in which consumption develops, either by changing its structure or by its volatility and growth.

Other economists, led by Larry Summers, are going to pitch in here. For Summers, secular stagnation is primarily the consequence of the weakness of aggregate demand. It can be explained by a change in savings behaviour and translates into a planned rise in savings before the adjustment of real interest rates. Ben Bernanke spoke of a "savings glut", but this glut was before the adjustment of real interest rates. Yet we can observe that this excess savings situation has pulled real rates so far downwards that they can go no lower and are actually becoming negative. In the end, this amounts to saying the economy will not manage to reach its potential growth, because it is not possible to lower real interest rates sufficiently. So we find ourselves caught in a "liquidity trap"! And in the end, if real interest rates can no longer be adjusted in order to balance savings and investment, the system will adjust itself towards an excess of savings.

Why such an excess? At first, Summers pointed to inefficient income distribution and little inclination for consumption among the richest. This would thus have contributed to a high accumulation of savings in developed countries. For Paul Krugman, this argument is not sufficient because savings rates actually fell in the United States between 2007 and 2017. However, Summers then clarifies his argument: the rise in savings would in fact be replaced by an increase in debt among the poorest households, and it is the slowdown forced by post-crisis debt which finally focussed attention on secular stagnation, which itself is very real.

Initially, the debate only seemed to concern America. But what matters however, as Eichengreen reminds us, is the level of savings worldwide! The reallocation of wealth—led by China and the oil-exporting countries—between developed countries and emerging ones (the latter with high savings rates designed to compensate for minimal social security) would thus have led to the observed rise of savings worldwide. According to Oliver Blanchard (Blanchard et al. 2014), during the 2000s, the savings rate of emerging countries rose by 10 points and would thus have contributed to a rise in worldwide rates of up to 1.7 points between 2000 and 2007. There is a wider concept of secular stagnation today as a result. It includes Japan, of course, stuck with sluggish growth since the 1990s and affected by serious demographic ageing; it also includes a Eurozone marked by low levels of recovery, inflation and investment. For Nicholas Crafts (2014), moreover, the risks of secular stagnation would in fact be much stronger for the Eurozone than for the United States, in particular, because of less favourable demographics, lower productivity growth and restrictive economic policies in force in the Eurozone. Some strongly support supply, others demand. Finally, a third school of thought has emerged: that of the hysteretic effect, which transforms something transitory into a permanent phenomenon. A hysteretic effect would be, for example, the loss of human capital linked to the persistence of long-term unemployment. Or it could be excessive prudence which leads businesses and households to accumulate precautionary savings in the form of liquid assets, to the detriment of investment, thus impeding future productivity. Hysteretic phenomena, which anticipate a lower potential supply over the long term, are also found in the effects of anticipation, thus impeding demand today and in future. Tomorrow's stagnation would largely be the fruit of our current behaviour, caused by doubts and fears in the face of so much uncertainty. It is a fascinating, multifaceted debate for economists troubled by their difficulty in understanding this highly complex period.

But we know that the very concept of stagnation does not belong exclusively to economists. It has always been at the heart of Western thinking about progress: those two hundred years of progressive ideology, born with enlightenment thinkers like Condorcet, for whom history is written in a kind of irreversible sequence towards "a progress of the human spirit"; or like Voltaire, who when comparing Rome and England, said: "the civil wars of Rome ended in slavery and those of the English in liberty" (Voltaire 1733). It is an "ideology" proclaimed by Leibnitz which spread during the nineteenth century with evolutionary theories, above all with Darwin's.

But this ideology, so widely shared by the West, the "history as an arrow" model, is confronted with reality and the expression of fears associated with badly managed progress. At the end of the nineteenth century, which could be thought of as peaceful and facing towards the future, certain voices were raised, saying they feared the worst: decline. The same is true of the English fear in 1860 of seeing the end of coal as a source of energy. It is also true of France, caught in the turmoil of a declining population which prevents it from joining the industrial revolution on an equal footing. The solution will certainly come from immigration, but in 1901, Edmond Théry was concerned about the rise in power of Japan and China: "The yellow peril which threatens Europe can be defined as follows: violent disruption of the international equilibrium on which the social order of the great industrial nations of Europe is built, a disruption provoked by sudden, abnormal and unlimited competition from a huge country" (Théry 1901).

We can see that societal change has often been experienced as negative. It was experienced as disorder: chaos threatening to destroy an older equilibrium. And it was ever thus, since we can find it mentioned in Ovid, or in Hesiod in *Works and Days*. After the ages of gold, silver and bronze, mankind lives in the age of iron, of labour, right up to Bossuet, who prefers to believe in the periodic decline of civilisations. It is therefore a cyclical concept of history which recalls the comings and goings of our societies, with no discernible continuous progress for humanity. Jean Gimpel reminds the West: "Technological progress is cyclical, as is most of history. The West has been privileged to live through two major cycles…within a civilization that has lasted now for a thousand years…But today the West has no new young nation in reserve and that momentum cannot be maintained" (Gimpel 1977). He describes the three centuries from the sixteenth to the eighteenth as a machine civilisation because of its dams, wind and water mills, rock mining industry, the agricultural revolution with three-year crop rotation and the Cistercians' model farms.

Georges Duby follows a similar path. "Between the year 1000 and the thirteenth century, society was carried along by tremendous material progress, comparable to that unleashed in the eighteenth century and which continues today" (Duby 1997). Agricultural production actually increased by five or six times in two centuries; the movement of people tripled, and people and goods moved faster. The most significant result of progress was the rebirth of towns. Then the Western world re-entered a phase of stagnation, which was secular because it lasted until the middle of the eighteenth century. "Thus, there was no noticeable progress in transport between the reign of Phillipe Auguste (1180-1223) and Louis XIV (1643-1715). The travelling time from Marseille to Paris remained largely the same for five centuries" (Duby 1997).

Let us come back to today's economists, who are plunged into doubt. What can they say with certainty? Quite simply that the combination of the effects of the slowdown in productivity gains, and of a demand not yet expressed via the new forms of consumption, is creating the elements of a difficult transition. And we use the term "transition" because, like all technical systems, the future, which is obviously based on many other changes besides those linked to digital technology, will take time to establish itself. This was the case with the first two industrial revolutions, which really emerged at the moment when Schumpeter's cluster of innovations enabled profound changes in production processes and in the nature of goods and services consumed. It is the convergence of these two phenomena alone which signals the end of the transition period and the restarting of a new growth.

What do we conclude from this? Certainly that the decades following the Second World War were a rare, if not unique period: the moment when the West succeeded in channelling technical progress and transforming it into social progress. That ended in the 1980s; then came the emerging countries' years of growth, with a proportion of humanity coming out of poverty. But history stops there today, since we do not know what the pace or the effects of the current technological innovations will be. Eventually, there will much optimism. But today, a real clear-headedness dominates. Even more so as we are only seeing the first signs of the transformations to come, which, we must remember, will involve all branches of science and technology.

What we do know is that we are in a period of uncertainty. We are far from having built the components of a new economic structure and new forms of consumption and production, i.e. the development of a real technological industrial revolution. That transition could take us down different paths: enslavement to new technologies, or prosperity founded on the new development of humanity.

## 2.2 Transition, the Precursor of an Industrial Revolution

Reports follow one after the other, all with a constant theme: all fields of human activity will be heavily disrupted by digital technology, bringing a worldwide change in the employment market. But this idea may not merit the vast amount written about it. The more these studies predict unmitigated disaster, the more they seem to be taken seriously; but today we can only observe one, single, fundamental change: the development of widespread disintermediation of marketing in most sectors. This means we bid farewell to traditional agencies for marketing, accommodation, travel, wholesale and retail business. But every industrial revolution, every new trajectory of the world economy, every new form of growth assumes that the standard of consumption will be profoundly changed. However, that is not yet the case today. These marketing shocks actually concern traditional goods and services. And it is here that we detect the very embryonic characteristics of this new world. The standard of consumption cannot change unless goods and services evolve in many other sectors than those involved today, for example, in energy, health, space exploration and the transmission of knowledge. What will the new substance of this revolution be? The entire fields of science, technology and innovation are at stake here, taking us well beyond a purely "digital revolution". This upheaval is not on its way today, as demonstrated by the slowdown in productivity and the still very marginal character of jobs created and value added from this new sector.

Can we, therefore, talk about an industrial revolution with equivalent effects to those seen on two occasions, at the end of the eighteenth century and in the second half of the nineteenth century? Let us first of all remember that Schumpeter qualifies the expression "industrial revolution" as "unfortunate"—which could, at first, seem paradoxical in relation to his theory on "creative destruction" but is not at all: "if we survey the course of economic history, we do not find any sudden ruptures, only a slow and continual evolution" (Schumpeter 1946). In fact, following the form of Kondratieff's cycles here, this great economist thinks that this "moment" is only one of the phases of the process of industrialisation and that it is an inherent part of capitalism.

"The fundamental impulse that sets and keeps the capitalist engine in motion comes from new consumer goods, new methods of production or transportation, new markets, new forms of industrial organisation that capitalist enterprise creates. […] The history of the apparatus of power production from the overshot water wheel to the modern power plant, or the history of transportation from the mail coach to the airplane. The opening up of new markets, foreign or domestic, and the organizational development from the craft shop and factory to such concerns as U.S. Steel illustrate the same process of industrial mutation – if I may use that biological term – that incessantly revolutionizes the economic structure *from within*, incessantly destroying the old one, incessantly creating a new one. This process of Creative Destruction is the essential fact about capitalism. It is what capitalism consists in and what every capitalist concern has got to live in" (Schumpeter 1942). This long citation is meant to remind everyone of the concept of time: of the long term, which is the only way to understand what a real transformation in the world economy is.

But let us return to those historic disruptions (Lorenzi and Bourlès 1994), even if they took time to establish themselves. "It will just not do to say that the horse-collar […] 'progressively reduced man's slavery'. […] Nor did the centre-line rudder […] pave the way for, and then ensure the success of the great maritime discoveries" (Braudel 1979). That warning from Fremand Braudel, despite his very ancient examples, illustrates the incorrect readings that have been made of the history of technology and innovations. We could say the same about the extrapolations made by futurologists of scenarios which could be brought about by the ongoing revolution in new information and communication technologies (NICTs). The sometimes grotesque and often incorrect parallels drawn between technical progress and past or future societies reveal, most of the time, a sort of narrow determinism. They also skip over the uneasy and usually long transitions from one epoch to another. Is it a trick of history or language which makes us borrow the term "epoch" from the ancient Greek term *epoché*, which means "stop" or "interruption" (indeed "suspended" for the sceptics)?

The lesson to draw from our argument is as follows: a great invention—let's take printing as an example—does not initiate a "revolution", a paradigm shift in Western society. Moreover, printing takes a long time to spread through society: about a century for the most educated at the time, and a lot longer to nurture a real revolution in thought: a new era of the life of the mind such as the Scottish, English and French Enlightenment or the German Aufklarung. The humanities and thus humanism may have found their father figure in Montaigne, but they were brought to maturity by the philosophers of the eighteenth century. And it is they, not technical innovation, who got history back on the march, under the banner of human progress.

We must never forget the significant effect of time in all this. The term "revolution" certainly evokes speed and disruption. But while disruption may be certain, speed is less so. The spread of the first industrial revolution, between 1785 and 1849, corresponds to just under three generations, which means change is both abrupt and slow. While the reality of an industrial revolution may be almost incontestable where the last two centuries are concerned, we still need to understand what it is and to pinpoint its key components. Invention does not, at first glance, belong to the field of economics. The discovery of a principle enriches the sphere of knowledge, but could stay confined there and not be translated into the economic sphere. Innovation, on the other hand, is an economic application and creates a new production function with a new method of using resources. Industrial revolutions are thus the expression of a rebalancing of technology which initiates new economic growth. If scientific research is continuous and if the same could be said for the rate of new inventions, there will only be certain periods when changes create a new technological balance, which itself brings a new structure of production and consumption, and therefore a new economic and social structure.

Two conclusions present themselves. An industrial revolution is not the result of a major innovation, but much more that of a "cluster of innovations" as Schumpeter said. This being the implementation of a technical system made up of linked innovations coming together, which puts more emphasis on the relationships between techniques than on the techniques taken individually. The second characteristic of an industrial revolution is that it concerns the nature of the goods and services consumed as much as the manner in which these are produced. This means a shock to our ways of life, the way we consume, our social model and our production techniques. Is this the case today, in what is known as "digital development"? Definitely yes, eventually. The issue is not the astounding onward march of digital technology, but rather knowing at what point it is going to transform into an industrial revolution. Let us quickly review its different elements. We see low productivity gains in the development of productivity per head in all major nations and in the development of multifactorial productivity. As a reminder, multifactorial productivity, which is also called "total factor productivity", represents the relative increase in growth which is not explained by the increase in labour and capital. This indicator is perceived as the principle vector of growth and therefore of technical progress. As the OECD says, variations in multifactorial productivity bear witness to the repercussions of developments in management practices, changes to commercial brands and the restructuring and development of knowledge in general, but also to network effects, the fallout from production factors, adjustment costs, economies of scale and imperfect competition and measurement errors. We can clearly see in the figures below the significant weakening of productivity gains, and that is over a relatively short period (Table 2.1).

But we wanted to go further, to observe what would happen if this tendency continued. Starting from the observed historical data, we have projected the indicator of multifactorial productivity according to three scenarios and applied this exercise to various developed countries: the United States, the UK, Germany, France, Italy, Spain and Japan. The

**Table 2.1**  Mean growth rates of multifactorial productivity in %

| Period | United States | UK | Germany | France | Italy | Spain | Japan |
|---|---|---|---|---|---|---|---|
| 1985–1994 | 0.8 | 1.1 | 1.7 | 1.4 | 1.2 | 1.1 | 2.1 |
| 1995–2004 | 1.4 | 1.6 | 1.0 | 1.2 | 0.2 | −0.2 | 0.7 |
| 2005–2015[a] | 0.6 | 0.2 | 0.6 | 0.3 | −0.3 | 0.0 | 0.5 |

[a]Data up to 2014 for Spain and Japan
*Source* OECD and the authors

first scenario assumes that the development of multifactorial productivity will follow the same development between 2017 and 2040 as that observed between 1985 and 1994. The second scenario assumes a similar development to that observed during the period 1995–2004. Finally, the third and last scenario assumes a development similar to the rates of growth observed between 2005 and 2015.

What do we learn from these graphs, which are only a simple simulation? We observe at which point trends can differ between, on one hand, The United States and The UK, and on the other, Germany, France and Italy.

We can see from these data that in the major developed economies, multifactorial productivity tends to stagnate, indeed slow down, in the scenario where it follows the most recent developments in growth, corresponding to a previously mentioned stagnation scenario (Graphs 2.1, 2.2, 2.3, 2.4).

- See the appendices for graphs for Spain, Italy and Japan.

Of course, all of this is of only illustrative value, because we can assume this only describes the transition phase in which we find ourselves. This is even more true because, in fact, productivity gains are extremely strong in the NITC sector. But there is the paradox: the size of the new technology sector, taken in the strict sense, is very small. This means that whatever its value as a driving force may be, it cannot constitute the new growth on its own.

The proportion of value added, defined as the contribution of labour and capital to production by this same sector to the total value added

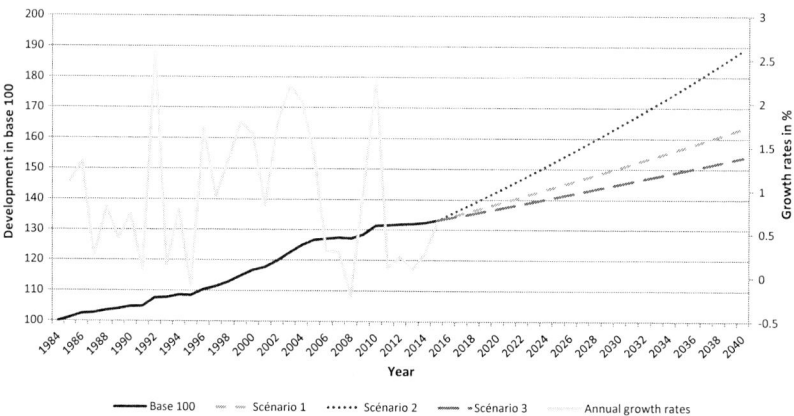

**Graph 2.1**  Total factor productivity, growth rates and development in base 100 from 1985: United States (*Source* OECD and the authors)

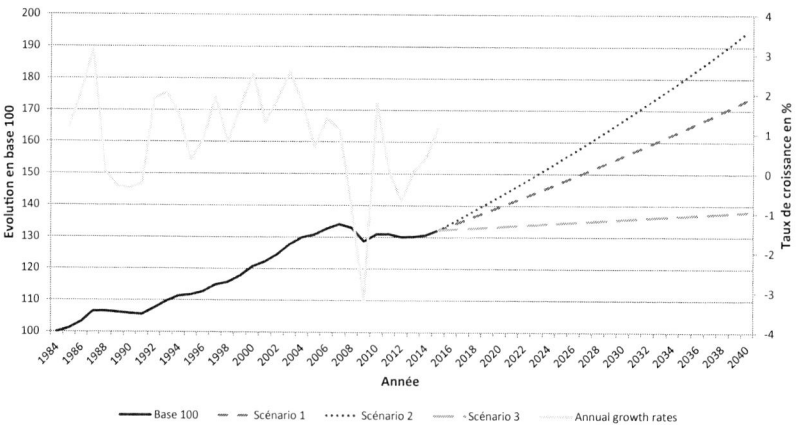

**Graph 2.2**  Total factor productivity, growth rates and development in base 100 from 1985: United Kingdom (*Source* OECD and the authors)

in the economy, seems to have stagnated since the end of the 1990s (Tables 2.2 and 2.3).

But what is more significant is the proportion of employment in NITCs: tiny today, even smaller in future. To illustrate this, we have taken three scenarios, simply to show orders of magnitude. First of all,

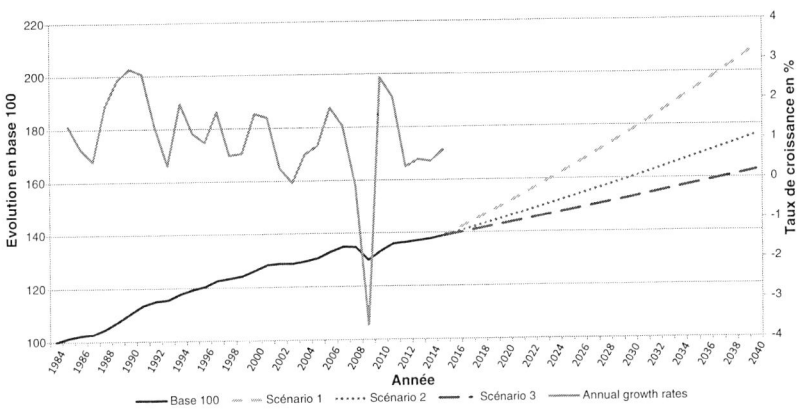

**Graph 2.3** Total factor productivity, growth rates and development in base 100 from 1985: Germany (*Source* OECD and the authors)

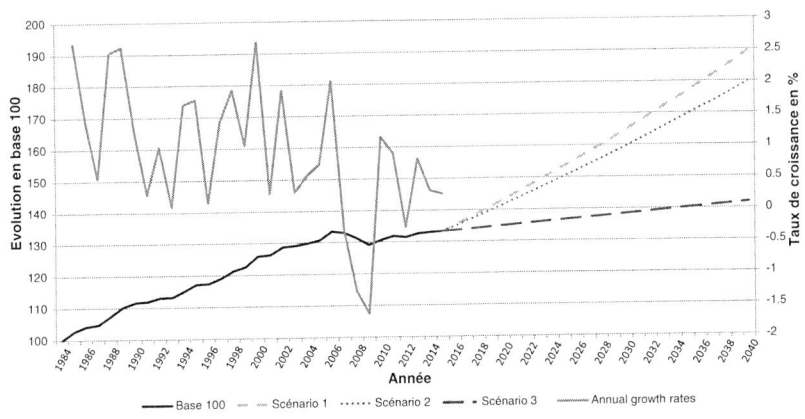

**Graph 2.4** Total factor productivity, growth rates and development in base 100 from 1985: France (*Source* OECD and the authors. See some more graphs in appendix)

we assume the productivity of the NITC sector follows the same trend as that observed between 1985 and 1995, and that the same change in the proportion of ITC employment is then applied to the period 2015–2040 (scenario 1). Secondly, we assume the productivity of TICs follows the same trend as it did between 1995 and 2005. So once

Table 2.2  Mean proportion of value added from the ITC sector expressed in %

| Period | United States | UK | Germany | France | Italy | Spain | Japan |
|---|---|---|---|---|---|---|---|
| 1997–2004 | 5.7 | 6.1 | 4.6 | 5.3 | 4.3 | 4.5 | 5.0 |
| 2005–2015[a] | 6.0 | 6.2 | 4.7 | 5.1 | 4.1 | 4.4 | 5.5 |

[a]Data up to 2014 for the United States
*Source* OECD and the authors

Table 2.3  Mean growth rates of value added by TICs expressed in %

| Period | United States | UK | Germany | France | Italy | Spain | Japan |
|---|---|---|---|---|---|---|---|
| 1997–2004 | 6.8 | 9.1 | 6.8 | 7.0 | 7.0 | 6.0 | 6.1 |
| 2005–2015[a] | 4.4 | 3.1 | 4.8 | 3.1 | 0.7 | 3.5 | 1.2 |

[a]Data up to 2014 for the United States
*Source* OECD and the authors

again we apply the same trend to the proportion of TIC employment between 2015 and 2040 (scenario 2). Finally, we imagine that the development in productivity of NITC sector businesses follows the trend of global productivity. In that case, the proportion of NITC employment remains unchanged compared to 2013 (scenario 3) (Table 2.4).

The results obtained show that in each individual case, the solution to employment problems is not to be found in the NITC sector. Is this proof that the industrial revolution is not actually miraculous? No, but it is an indication that the rather naïve picture presented of the NITC sector is a long way from reality.

Well beyond all of this, the debate rages on over productivity gains, as seen today and in the past. Some economic historians believe that in the short term, a great invention initially tends to reduce productivity, not increase it. Paul David of Stanford University illustrates this concept using the example of electricity (David 1990). Before the introduction of electric motors in factories, machines were powered by steam engines. However, the appearance of electricity and the self-contained electric motor necessitated a total restructuring of labour, which took some time. Ultimately, the time taken to implement that reorganisation would have disrupted production and led to a lowering of productivity.

**Table 2.4** Development of ITC sector employment, expressed as a % of total employment

|  |  |  |  | Scenario 1 | Scenario 2 | Scenario 3 |
|---|---|---|---|---|---|---|
|  | 2001 | 2007 | 2013 | 2040 | 2040 | 2040 |
| United States | 4.1 | 3.4 | 3.3 | 1 | 3.3 | 3.3 |
| UK | 3.6 | 3.4 | 3.3 | 2.5 | 2.9 | 3.3 |
| Germany | 2.8 | 2.9 | 2.8 | 3.2 | 2.5 | 2.8 |
| France | 2.9 | 2.7 | 2.7 | 2 | 2.6 | 2.7 |
| Italy | 2.5 | 2.5 | 2.5 | 2.4 | 2.8 | 2.5 |
| Spain | 1.9 | 1.7 | 2.1 | 1.1 | 3.7 | 2.1 |
| Japan | 2.3 | 2.6 | 2.6 | 3.8 | 2.5 | 2.6 |

*Source* OECD and the authors

This argument is contradicted by Summers. If the drop in productivity we see now was the result of a restructuring of labour equivalent to that caused by electricity, this phase would be accompanied by the creation of jobs to support businesses in their restructuring! Eichengreen's response to this is that unlike the introduction of electricity, the establishment of new technologies does not create significant demand for labour, compared to the number of jobs likely to disappear as a result.

We can see that no single theory prevails. But the arguments exchanged even touch on the methods of calculation. Some believe we are in an abundance economy, which explains the refusal to pay for digital products and services and prevents these new openings from being included in measurement of GDP or of productivity. How to account for goods which do not have a market value? Hal Varian takes the example (Varian 2016) of digital content financed by advertising: an economic model which is widely exploited on the Internet. He states that the National Economic Accounts of the US Bureau of Economic Analysis do not include them in the calculation of GDP because advertising is considered a marketing cost. Thus, a content provider who goes from a pay-per-view economic model to a model financed by advertising contributes to the lowering of national GDP. The same is true of photographs, yet the number taken has multiplied by 20 in the space of fifteen years, from 80 billion in 2000 to more than 1500 billion in 2015.

Their unit cost has gone from 50 Euro cents to zero, particularly with the development of smartphone cameras. But since they are neither bought nor sold, they do not appear in the GDP calculations.

The techno-optimists denounce official statistics which do not take account of new technologies and the improvements in quality of life which they generate. Yet, by the same token, do we not underestimate, in a much more significant way, the time spent working? Growth in productivity does correspond to the increase in production per unit of labour. Yet, for most people working in the service sector, the fact that it is possible to be connected all the time, particularly outside of work, is a reality which also escapes the official statistics. And in this case, the fact of not being able to account for that work time leads to an overestimation of the level of productivity.

The conclusion to this fascinating debate, from our point of view, comes back to Gordon. According to him, productivity began to flag well before the appearance of mobile phones and the Internet. Even more importantly, he believes GDP has always been underestimated. When Henry Ford dropped the price of his Model T from 900 dollars in 1910 to 265 dollars in 1923, while at the same time improving the quality, the consumer price index did not include cars in its calculations (Gordon 2014). Similarly, if the large drop in infant mortality at the beginning of the twentieth century had been taken into account in the calculation of GDP, it would have had a more significant effect on GDP than all measured consumption (Nordhaus 2003).

The question is therefore far from decided. But the facts are these: if the world economic slowdown really is happening, it is partly explained by the difficulty in improving the efficiency of production systems as we have in the past. That is the strictly economic version. However, that analysis is not sufficient, because we are seeing a profound transformation in our social structures, expressed and developed via networks, the real significance of which we are still struggling to grasp today.

## 2.3   A Decisive and Growing Influence

Social networks, which emerged with the Internet, have mostly been heralded as an almost insuperable stage in the emancipation of the individual from everything that could be controlled by authority. No more restrictive hierarchy; no more straightjacket of Ford-style business. We have reached the age of conviviality and sharing: a virtual society of equals or peers who short-circuit traditional top-down communication and have even less time for relationships based on power and domination. This refrain lingers on, despite the sharpest critics emerging here and there since the mid-1990s and the advent of widespread connectivity. However, this heralding of a new world, perhaps the best of worlds, struggles to turn its ideas into reality. Jacques Ellul, that harsh critic of technology, in the etymological sense of a discussion of technical methods, would without any doubt have once again called the bluff of this utopian ideology, and repeated his warning about "the fascinated man".

We have to look further. We cannot reduce the expression "social networks" just to Facebook, Twitter, Instagram, LinkedIn or other digital accounts: a world proclaimed to be peaceful and open. The French term for "network" (réseau) has a history which tells us much about its meaning, as rich as it is ambivalent. It appeared from 1694 onwards, in the Académie Française Dictionary, to indicate a creation of silk and threads that weavers call an "entrelacement" (interlacing) which is the way to arrange and "regulate" the threads. In the nineteenth century, the word arrived at the right moment for medicine to use it to describe the circulatory system and the nervous system, and by extension in the twentieth century, road and rail traffic. Complementary and sometimes contradictory concepts of circulation and topology, but also of control and cohesion, live together under this one term.

What can we conclude from these multiple meanings, apart from the fact that they are very far from the current perception of "social networks"—a perception which skips over the shadowy areas exposed and analysed so well by Georg Simmel. For this German philosopher and sociologist, who often features in the genealogy of the expression "social network", it is not so much individuals or society which counts, but that area in between: the relationship, the interaction between individuals. Sociology is thus "the science of the forms of reciprocal action". By using the term "form", we avoid thinking in terms of the nature of relationships (friendship, business, etc.) and focus on their methods, such as domination, imitation, competition and conflict. A task which, according to Simmel, enables the creation of "a geometry of the social world". Such is the case with the poor: an example he likes so much, he dedicated a book to it: "The poor person, sociologically speaking, is the individual who receives assistance" (Schermer and Jary 2013).

This reading undoubtedly enables us to place current social networks within a wider issue than the heavily commented-on question of the NITCs. It also gives us a way out of the dominant ideology about the "moment", experienced as a major revolution seeming to come from nowhere, with no precedent. We could say, following Simmel here, that social networks have their place in those "reciprocal actions which support all the firmness and flexibility, multiplicity and unity of life in society" (Simmel 1911).

And yet, economics has appropriated the subject with a quite simple and well-executed logic (Colin et al. 2015), which is the illustration of a digital economy which expands via growing returns. As William Brian Arthur (1996) shows, increasing the number of clients leads to a better quality of service, which translates into direct and indirect network effects. This is what Metcalfe's Law summarises; it determines the value of a telecommunications network as proportional to the square of the number of connected users. For a long time, this whole idea was more intuitive than proven. It was only in July 2013 that researchers from the Netherlands succeeded in analysing European profiles of Internet use for a long enough period and found a proportionality of $n^2$ for smaller values of $n$ and a proportionality of $(n \times \log n)$ for large values of $n$ (Madureira et al. 2013). Several demonstrations of the validity of Metcalfe's law have been carried out for social networks. Metcalfe himself offered a model using Facebook data over ten years with a proportionality of $n^2$. Let us look ahead to 2030 to get an idea of the strength of these networks. If we imagine what the number of Facebook users would be, we can see what impressive figures we get by applying Metcalfe's Law: almost the entire population of the world would be connected to the network by then (Graph 2.5).

We know there is an enormous difference between current thinking about the power of new technology today and the more modest reality, but the power of that thinking depends on networks. We must first revisit the ideas of leading figures in genetics, new energy, nanotechnology and space technology. And incidentally this concerns major industrialised nations and emerging countries equally, as shown by the emerging movement of sector leaders in China.

The Chinese "BATX" is a very enlightening example. The acronym refers to the Chinese digital giants Baidu, Alibaba, Tencent and Xiaomi, often compared to the American "GAFA".[2] The GAFA firms have very little presence in China, because the digital economy is tightly controlled by the authorities. This has benefited local firms and enabled China to create its own digital ecosystem.

We can find endless similar examples of this current thinking. They form the basis of the creation of strategies for network use and of a growing power, based on growing returns, at least for the NICTs. In this instance, Jeff Bezos, founder of Amazon and Blue Origin, declared that his vision of the future: "millions of people living and working in space" and the New Glenn launch vehicle were a "very important step"[3] in that direction. New Glenn aims to compete with SpaceX in order to

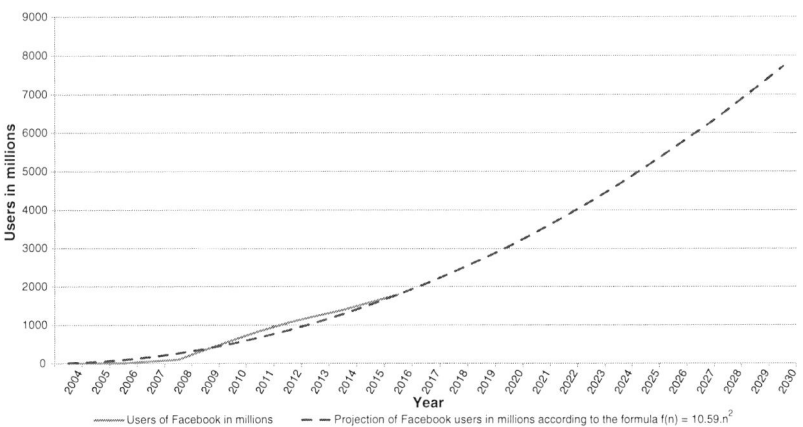

**Graph 2.5** The application of Metcalfe's Law to Facebook (*Source* Facebook [https://investor.fb.com/investor-news/] and the authors)

make space accessible by drastically reducing launch costs. Ray Kurzweil is also a leading, regular exponent of forward-looking statements about new technology in its widest sense, particularly on genetics. "By understanding the information processes underlying life, we are starting to learn to reprogram our biology to achieve the virtual elimination of disease, dramatic expansion of human potential, and radical life extension", he writes in his book *The Singularity Is Near* (Kurzweil 2006). In the same book, he cites Horst Störmer, for whom: "Nanotechnology has given us the tools…to play with…atoms and molecules. Everything is made from it… The creative possibilities appear to be unlimited".

But, beyond Metcalfe's Law, we must question the reason for the growing returns. Where do they come from? In reality, digital technology makes it easier to verify the other party in a transaction, check their reputation, communicate easily and keep track of exchanges: in other words, to build trust between parties who do not know each other (Dyer and Wujin 2003). This has led to the appearance of platforms where amateurs and semi-professionals can pitch to clients in ideal, safe conditions and offer services which are sometimes better quality than those offered by traditional providers. Growing returns are also generated by machine learning. It is used by digital businesses to constantly improve their performance (costs, efficiency, quality) via the collection and processing of massive flows of data. These networks and their power have much more far-reaching consequences than a simple improvement in the efficiency of services. In fact, the diminishing costs of those services are giving rise to a race towards concentration. In most cases, the market then becomes dominated by the firm which manages to harness an exponential growth before the rest, propelled by a snowball effect. In fact, it is not enough to be the first business to enter the market, but rather the one with the earliest and most sustained growth, a victory based on "winner takes all" (Kutcher et al. 2014).

This concentration, which leads to monopolisation, is perfectly illustrated by the dizzying trajectory of the large digital companies. Their exceptionally high valuations enable them to build concentration strategies. In fact, the markets they operate in are characterised by the undivided dominance of whichever company achieved exceptional growth before the others. We must draw attention to this process, because the company which acts soonest and is most dynamic wins: competitors subsist and monopolies may not last as long as in the traditional economy. This is actually explained by the intensity of the competition, enabled by low entry costs and by dependence on users. Economies of scale and network effects actually come from the trust inspired in users. Faced with the introduction of competition by consumers, the digital giants can no longer hide forever behind high-cost infrastructure or regulatory barriers.

Nevertheless, the new economy stands out because of the speed of its consolidation of the relationship between finance and the real economy. The digital giants are implementing a strategy to control the sector, as demonstrated by large service companies such as Uber, which distributes its app via Google, Apple and Facebook, depends on Google Maps for geolocation, uses Google and Apple payment systems and Amazon's technical infrastructure. This type of company, dubbed a "unicorn", could not have developed so quickly without using the services of the digital giants, and this allows the giants to control the unicorns from a distance without having to buy them.

On a more general note, the digital giants finance their own development and that of the sector. They fund their own innovation, investing a major amount of their cash in R&D, reinforcing their position as market leaders. Furthermore, they devote colossal sums to acquiring start-ups and other digital businesses. Digital giants also fund numerous NITC-related companies in many fields: unicorns, fintech, biotech, cars of the future, etc. Investment in fintechs, for example, is at the heart of the new economy, because they are the textbook example of start-ups; they are making profound changes to the banking and finance sector and are widely funded by stakeholders with a means of controlling the competition that they can create. And what is true for fintechs is, or will be, true for all sectors of the new economy.

But the worst is never guaranteed to happen. The current concentration and economic power of a small number of companies are going to grow until a political, anti-monopoly reaction may occur, probably beginning in the United States. Admittedly, up to now, the financing of the new economy is the main tool of this growing power, of this concentration, whether this funding flows directly via takeovers or indirectly via funds which end up selling the shares they have acquired to the major players.

These networks and those who created them have understood their importance very well, since—must we remind ourselves—Facebook is in size, equivalent to the leading country in the world. They create their power in complete obscurity, based on a prophetic vision of what a world in which they are the leading institutions should be. For Larry Page, co-founder of Google, his company "must create its own future, because in technology we need revolutionary, not incremental change" (Page 2014). But not everything is as simple as it is today.

"To develop and impose its vision of progress, this elite must now get involved in the political agenda. It knows it has enormous power. So, if it wants, for example, driving licences for its driverless cars, it will need to convince the American and other state regulators, which will certainly not want them", explains Frédéric Martel (Auciello 2014). After finance, comes politics. And whether we want it or not, the two could end up in conflict.

## Notes

1. Potential growth is an estimate of growth rates when factors of production are used to their full capacity.
2. Acronym referring to Google, Apple, Facebook and Amazon.
3. Mosher D (2018) Jeff Bezos Says That Amazon Is Not His "Most Important Work". Available via http://uk.businessinsider.com/jeff-bezos-blue-origin-rocket-company-most-important-2018-4.

## Appendix

See Graphs 2.6, 2.7, 2.8.

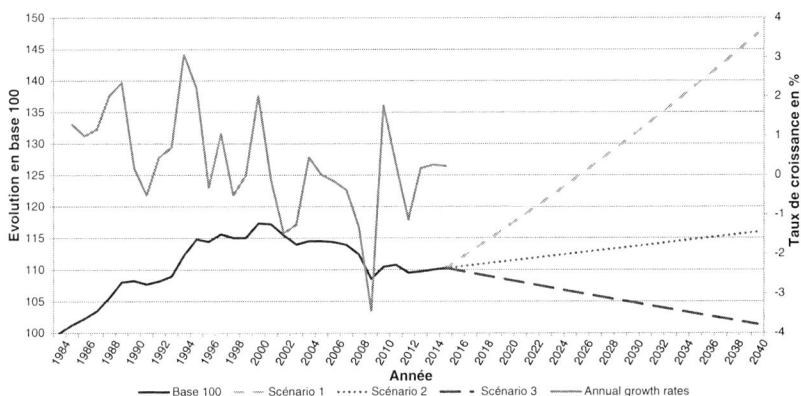

**Graph 2.6**   Total factor productivity, growth rates and development in base 100 from 1985: Italy (*Source* OECD and the authors)

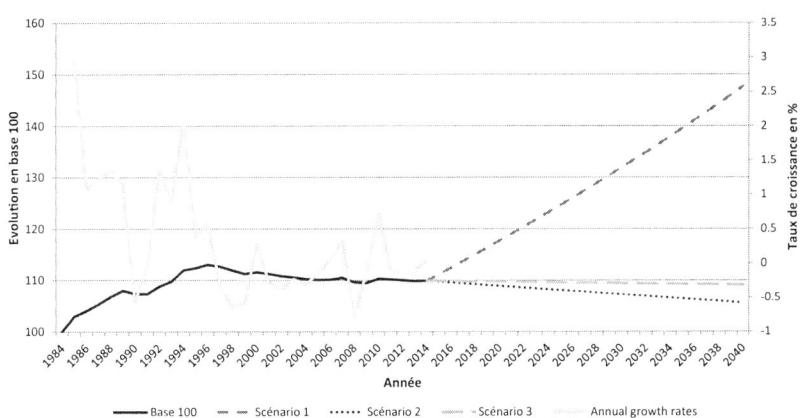

**Graph 2.7**   Total factor productivity, growth rates and development in base 100 from 1985: Spain (*Source* OECD and the authors)

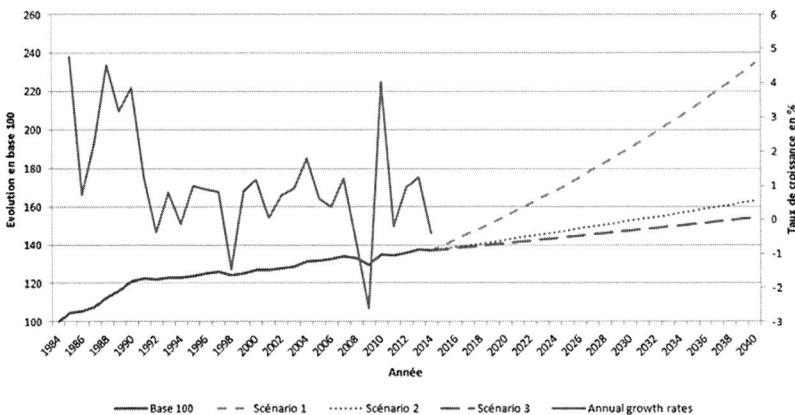

**Graph 2.8** Total factor productivity, growth rates and development in base 100 from 1985: Japan (*Source* OECD and the authors)

# References

Arthur, W. B. (1996). Increasing Returns and the New World of Business. *Harvard Business Review, 74,* 100.

Auciello, D. (2014). *Ces géants mégalos qui dirigent le monde* [Those Megalomaniac Giants Who Lead the World]. Bilan.ch. Available via https://www.bilan.ch/economie/ces_geants_megalos_qui_dirigent_le_monde.

Blanchard, O., Furceri, D., & Pescatori, A. (2014). A Prolonged Period of Low Real Interest Rates? In *Secular Stagnation: Facts, Causes and Cures.* Center for Economic Policy Research Press. Available at VOX CEPR Policy Portal https://voxeu.org/content/secular-stagnation-facts-causes-and-cures.

Braudel, F. (1979). *Civilization and Capitalism 15–18th Century, Vol 1: The Structures of Everyday Life: The Limits of the Possible* (2002 ed., S. Reynolds, Trans.). London: Phoenix Press.

Colin, N., Landier, A., Mohnen, P., & Perrot, A. (2015). Economie numerique. *Les Notes du Conseil d'analyse économique* (26). Available in English as The Digital Economy, via http://www.cae-eco.fr/IMG/pdf/cae-note026-en.pdf.

Crafts, N. (2014). Secular Stagnation: US Hypochondria, European Disease? In *Secular Stagnation: Facts, Causes and Cures.* Center for Economic Policy Research Press. Available at VOX CEPR Policy Portal https://voxeu.org/content/secular-stagnation-facts-causes-and-cures.

David, P. A. (1990). The Dynamo and the Computer: An Historical Perspective on the Modern Productivity Paradox. *Papers and Proceedings of the Hundred and Second Annual Meeting of the American Economic Association, 80*(2), 355–361.

Duby, G. (1997). *An 1000 – An 2000 – Sur les traces de nos peurs*. Paris: Éditions Textuel.

Dyer, J. H., & Wujin, C. (2003). The Role of Trustworthiness in Reducing Transaction Costs and Improving Performance: Empirical Evidence from the United States, Japan and Korea. *Organisational Science, 14*(1), 57–68.

Eichengreen, B. (2014). Secular Stagnation: A Review of the Issues. In *Secular Stagnation: Facts, Causes and Cures*. Center for Economic Policy Research Press. Available at VOX CEPR Policy Portal https://voxeu.org/content/secular-stagnation-facts-causes-and-cures.

Gimpel, J. (1977). *The Medieval Machine: The Industrial Revolution of the Middle Ages*. New York: Penguin.

Gordon, R. J. (2012). *Is US Economic Growth Over? Faltering Innovation Confronts the Six Headwinds* (Working Paper 18315). National Bureau of Economic Research. Available via NBER https://www.nber.org/papers/w18315.pdf.

Gordon, R. J. (2014). The Turtle's Progress: Secular Stagnation Meets the Headwinds. In R. Baldwin & C. Teulings (Eds.), *Secular Stagnation: Facts, Causes and Cures*. Center for Economic Policy Research Press. Available at VOX CEPR Policy Portal https://voxeu.org/article/turtle-s-progress-secular-stagnation-meets-headwinds.

Hansen, A. (1939). Economic Progress and Declining Population Growth. *The American Economic Review, 29,* 1–15.

Keynes, J. M. (1937). Some Economic Consequences of a Declining Population. *The Galton Lecture, Eugenics Review, 29*(1), 13–17.

Kurzweil, R. (2006). *The Singularity Is Near*. London: Penguin.

Kutcher, E., Nottebohm, O., & Sprague, K. (2014). *Grow Fast or Die Slow*. McKinsey Global Institute. Available via https://www.mckinsey.com/industries/high-tech/our-insights/grow-fast-or-die-slow.

Lorenzi, J. H., & Berrebi, M. (2016). *A Violent World*. London: Palgrave.

Lorenzi, J. H., & Bourlès, J. (1994). *Le Choc du progrès technique*. Paris: Economica.

Madureira, A., den Hartog, F., Bouwman, H., & Baken, N. (2013). Empirical Validation of Metcalfe's Law: How Internet Usage Patterns Have Changed Over Time. *Information Economics and Policy, 25*(4), 246–256.

Mokyr, J. (2014). Secular Stagnation? Not in Your Life. In *Secular Stagnation: Facts, Causes and Cures*. Center for Economic Policy Research Press. Available at VOX CEPR Policy Portal https://voxeu.org/content/secular-stagnation-facts-causes-and-cures in the United States.

Nordhaus, W. D. (2003). The Health of Nations: The Contribution of Improved Health to Living Standards. In K. M. Murphy & R. H. Topel (Eds.), *Measuring the Gains from Medical Research: An Economic Approach* (pp. 9–40). Chicago: University of Chicago Press.

Page, L. (2014). *Where's Google Going Next?* Ted Conference. Available via https://www.ted.com/talks/larry_page_where_s_google_going_next.

Schermer, H., & Jary, D. (2013). The Poor. In *Form and Dialectic in Georg Simmel's Sociology*. London: Palgrave Macmillan. Available via https://link.springer.com/chapter/10.1057/9781137276025_5#citeas.

Schumpeter, J. A. (1942). *Capitalism, Socialism and Democracy* (p. 83). New York: Harper and Brothers.

Schumpeter, J. A. (1946). Capitalism. In R. V. Clemence (Ed.). (2003), *Essays on Entrepreneurs, Innovations, Business Cycles and the Evolution of Capitalism* (pp. 189–210). Piscataway, NJ: Transaction Publishers.

Simmel, G. (1911). Le domaine de la sociologie. In *Sociology and Epistemology*. Paris: PUF.

Sweezy, P. (1942). *The Theory of Capitalist Development*. New York: Monthly Review Press.

Théry, E. (1901). *Le Péril Jaune*. Paris: Félix Juven.

Varian, H. (2016). Intelligent Technology. *Finance and Development, 53*(3). Available via https://www.imf.org/external/pubs/ft/fandd/2016/09/varian.htm.

Voltaire. (1733). Letters on England. In *The Collected Words of Voltaire: The Complete Works* (Kindle ed.). New York: PergamonMedia.

# 3

## The High Tech Eden

Why refer to technological progress as Eden? Because the Garden of Eden evokes the Earthly paradise: a world of enjoyment, pleasures and delights… but it also refers to the episode where man and woman were marked out for wanting to appropriate the knowledge of good and evil, which led to The Fall. Could knowledge bring mankind back to that perfect world, that Earthly paradise, far from confusion and division and make them immortal again? Some "techno-prophets" believe it can.

There can be no doubt at all that since the mid-1940s, we have experienced an unprecedented acceleration in knowledge: from the infinitely big to the infinitely small. This exceptionally fast development in our understanding of matter questions again our relationship to living things and our approach to spiritual and religious life. Curiously, Norbert Wiener, the great mathematician and founding father of cybernetics (the revolutionary new cross-disciplinary science which emerged in the 1940s), asks himself about re-establishing links between science and religion in 1964, the very year of his death (Wiener 1964). By revisiting the Jewish legend of the Golem,[1] this visionary demonstrates that societies are constructed around shared values which weave the fabric of trust and belief that people consider central to their relationships

© The Author(s) 2019
J.-H. Lorenzi and M. Berrebi, *Progress or Freedom*,
https://doi.org/10.1007/978-3-030-19594-6_3

and cohesion at any given moment. This is the very crux of the matter. For the first time since the Enlightenment, a tremendous hope, which is naïve, irrational and therefore dangerous, is lighting up the world by declaring that science can overcome all human limitations.

We are embarking on a risky exercise: certainly questionable, but useful all the same. We are taking four so-called disruptive technologies, which represent profound changes in the value systems we live by and most importantly, give rise to insane prophetic statements about how the world will be. We want to use these examples to demonstrate the important nature of these innovations and the risks posed by the hazardous and false predictions they can elicit. None of these technologies alone can explain the terrific dilemma confronting us today, but each of them illustrates the economic and philosophical arguments that we must tackle.

The Bible recounts the first temptation of human omnipotence. Let's remind ourselves of the Tower of Babel and that community, all speaking the same language or code, which had acted on its desire to appear equal to God and who would be punished for having attempted to overreach the divine. Humans, or more accurately, the new techno-prophets, are taking up this position again and tell us we are capable of rebuilding trust, reshaping freedom and achieving perfection and immortality.

This could raise a smile, if these same men did not have such a large audience and such power. But the time has come to put innovation back in its correct place. The four values which we believe are fundamental to the coherence and acceptability of our democratic societies will be transformed by the arrival of these four technologies. Their sole purpose here is to illustrate the profound movement our societies will be confronted with in the next twenty years. Technology is never neutral.

## 3.1    Artificial Intelligence or Dehumanisation

Perhaps it is excessive to evoke the term "dehumanisation" in relation to what appears to all of us to be the most iconic technological development of the age. However, dehumanisation is closely linked to servitude and alienation, as we can surmise and will discuss later. Nevertheless, it is worth expounding on, because it involves a process of people seeing their dignity as human beings taken away, which is legal in modern societies. Etienne de La Boétie was steeped in ancient Greek texts when he wrote, while still a student, *De la servitude volontaire, ou le Contr'un* (The Discourse on Voluntary Servitude) a surprising text which won him the unconditional friendship of his junior, Montaigne: a text which scrutinises and dismantles what could be called the balance of terror. Confronted with the tyrant, who is himself made inhuman by declaring himself superman and imposing his subjective view as the objective truth, the governed are also inhuman, since: "it is the nations who are allowing themselves to be castigated". Which amounts to saying: the tyrant's power is not imposed by force and that people allow themselves to be abused by outward trappings rather than "desiring the liberty and reason" they are endowed with. Custom and habit, La Boétie argues, make them accomplices of their tormentor and numb their reason, so that people "promptly fall into such complete forgetfulness of their freedom that they can hardly be roused to the point of regaining it" (De La Boétie 1576). The paradox of voluntary servitude happens the way it does because it is a truly unbelievable relationship, which is, however, powerfully documented by the philosopher, who hates nothing more than power concentrated in one individual. Dehumanisation is no more than the proxy of the will of power: a state which people allow themselves to be drawn into through ignorance, but also by flattery, blackmail and indeed corruption, which tyrants are so skilled at exploiting. And because the subjects forget that the tyrant owes them his power, step by step (five or six ambitious people is enough at first) layer by layer, by "moral inertia", political evil spreads. People are no longer any more than "homunculi", "debased" and "stupefied". This charge levelled at voluntary tyranny and dehumanisation is all the more interesting because it uncovers the secret of "consenting" dehumanisation, which our intellect rails against, but is confronted with all the same. La Boétie has an intellectual courage which—and this may be too bold a leap for many—recalls in some ways the philosopher Simone Weil and her Factory diary (Weil 1941–1942). This dedicated philosopher, who already foresaw the Nazi peril in 1932, decided to adopt and study the worker's condition. From December 1934 to the summer of 1935, she worked as an apprentice on the Alstom assembly line. It was there she observed that "things play the role of people and people play the role of things: that is the root of evil". She experienced hunger, extreme fatigue and verbal abuse, yet she carried on, in order to measure the extent of dehumanisation at work in Fordist factories. People, as such, no longer have a place in that society, or if they do, only in a state of living death: not so distant from La Boétie's "dehumanised" people.

But can we link this fundamental fear as easily to today's technologies, which are still at the embryonic stage? There can be no doubt we can. Think back to the end of December 2016. A video goes viral on the Internet of a motorist, Hans Noordsij, going at full speed along the motorway in "autopilot" mode in his Tesla car (made by Elon Musk). To the driver's complete surprise, the Tesla brakes suddenly. We have to wait a few more seconds to understand why: another car appears, also approaching at great speed and collides violently with a third car, which was a few metres in front of Hans. The intelligent automatic pilot system of Hans' car had anticipated the accident. By braking a few seconds before, it enabled him to avoid a pile-up he could never have foreseen. This kind of feat, attributable to an application of artificial intelligence widely disseminated by Elon Musk's company, is presented as an unprecedented, qualitative leap forward, as a disruption, as though it were not the result of history, but had emerged straight out of a science fiction film.

History does not get a good press, as the history of science knows from harsh experience. Well before Ray Kurzweil, a group of exceptional scientists, somewhat forgotten today, met during the pivotal years of the 1950s at the Macy Conferences to map out the outlines of a new interdisciplinary science, Cybernetics, an: "entire field of control and communications theory, whether in the machine or in the animal" (Wiener 1948). This approach, intended to explore the realm of the mind using a scientific approach via the emerging fields of automation, electronics and mathematical information theory, made a big impact at the time. In *Le Monde* on 18 December 1948, Père Dubarle wrote: "A new science: Cybernetics. Towards the rule of the machine". He continued: "Will the mechanical manipulation of human relationships create the 'Brave New World' one day?" We can see how old this question is: just like its protagonists.

What we call "artificial intelligence", the current common expression for one of the most promising branches of human knowledge, varies according to different people's ideas. For some, artificial intelligence is already here, very much present, with Google and its semi-automatic search engine, Facebook and Amazon and their predictive algorithms, cars with driving aids and so on. For others, that path is still ahead of us. They believe artificial intelligence will turn our way of life completely upside-down, so that everyone will delegate all sorts of tasks and decisions to intelligent robots. How long before we live in the world portrayed by director Spike Jonze in his film *Her*: a world where machines have their own consciousness, where robots can think, reflect, debate, laugh and experience their own feelings?

Let's go back in time. Marvin Minsky, a very early enthusiast of electronics, science and science fiction, is a father figure to artificial intelligence. He was a doctor in mathematics who had the opportunity to exchange ideas at the Macy Conferences with that other scientific prodigy, John von Neumann, one of the founders of computer theory. He also rubbed shoulders with Alan Turing, another brilliant researcher, with whom he wrote a paper on the functioning of the brain and the nature of intelligence, comparing them to machines. In 1951, he built the first neural network simulator. He then joined MIT, where in 1959 he created the MIT Artificial Intelligence Project, later to become the Artificial Intelligence Lab, with his colleague John McCarthy. As for Alan Turing, he attempted to define a standard for considering a machine "conscious" through his famous "Turing Test" (Turing 1950). The test involves letting an adjudicator interact verbally with another person and with a machine. If the adjudicator cannot tell the machine from the person, we can conclude the machine is "conscious". Minsky's concerns about neural networks and the possible dead end of artificial intelligence, expressed in 1969 with his colleague Seymour A. Papert (Minsky and Papert 1969), brought about the so-called AI Winter, but this ended in the 1980s.

Experts today distinguish between three major types of artificial intelligence. The first only concerns limited artificial intelligence, or NAI.[2]

This actually describes all the intelligent robots that we know today, which are capable of completing a precise task with as much, indeed sometimes more, skill than a human being: the smart personal assistant, voice and image recognition, simulation software in video games, fraud detectors in financial transactions and so on. The most iconic examples are definitely super-computers.

IBM set itself the goal of building a computer capable of winning the American TV game show *Jeopardy!* When it won in 2011, "Watson" consulted almost 200 million pages per question in less than three seconds, evaluating the probability that each answer was correct. Google's computer, AlphaGo, achieved another success in 2016 by winning the game Go: a much more significant achievement than Deep Blue beating Garry Kasparov at chess in the 1990s. There are millions of possible combinations in the game of Go. Yet in the space of five months, AlphaGo taught itself to play. The machine got progressively better thanks to two learning techniques: "deep learning", which can be based on a "non-supervised" model, and "reinforcement learning", which is made up of repeated experiences. In short, AlphaGo learned to learn and in doing so, beat the South Korean, Lee Sedol, who was considered the best player in the world.

The second type of artificial intelligence is far more significant: strong, high-level or "general" artificial intelligence.[3] This describes a robot which resembles a person in that it can learn from lived experiences in various fields. We are not there yet. Finally, the ultimate phase will be artificial super-intelligence. This describes an artificial intelligence which surpasses humans in all fields. This would be the end of Homo sapiens and the gateway to a world like that of Fritz Lang's *Metropolis*, where humans are slaves to machines. But wasn't that also depicted in Chaplin's *Modern Times*, in factory assembly lines ruled by inhuman rhythms repeating the same, single gesture? Will it be the end of human dignity and immutable rights? Is it a return to Plato's cave?

Many people fully expect humans to be surpassed. The reigning heroes of Silicon Valley and new Russian tycoons like the magnate Dmitry Itskov spend fortunes on it, while research centres appeal to the generosity of the new princes: the 0.1% richest people on the planet. In his survey conducted on 550 experts (Müller and Bostrom 2016), the

Swedish philosopher Nick Bostrom, director of the Future of Humanity Institute at Oxford and known for his anthropogenic approach, revealed that half the experts questioned estimated there was a one in two chance of seeing strong artificial intelligence before 2040. But this quest for strong artificial intelligence is not without danger and could pose existential risks for humanity.

Robots may have immense capacities for computing and processing, but at the moment the human brain is still superior to a machine. It alone can learn as quickly as it does, in an unsupervised way. The human brain possesses that capacity for abstraction which enables it to manipulate data, but also concepts and symbols. All the same, the progress observed in robots' computational power does give us a glimpse of the appearance of a new era, called the "Cognitive Era". The Cognitive Era describes the exact moment when all computers will be able to learn independently from non-structured data and from their own experiences.

How do we push the boundaries of the computational power of traditional computers? The human brain functions around 10,000 times better in terms of energy efficiency than the best artificial intelligences. According to Mark Mills, computers today consume 10% of world electricity production. This means the exploitation of future digital data, due to multiply by a factor of 1000 during the next fifteen years, can only be done on condition that there is a revolution in energy efficiency. Among the possible answers, many count on the development of quantum computers. Some organisations, like IBM, suggest that to equal the human brain, we must take inspiration from it, including imitating its very structure. In the same way that a brain is fuelled by sugar for energy and blood for cooling its different areas, organising computer components in a 3D matrix similar to a human brain could make it possible to use its cooling fluid to provide energy to chips. The fluid, as well as dispersing heat, could also fuel an electronic system to supply electricity to processors. This system, inspired by the human brain, could improve energy efficiency and computational power.

In a paradox of paradoxes, our technology prophets are the ones warning us against the risks involved. Elon Musk, Bill Gates, Stephen Hawking, Steve Wozniak—numerous experts have declared themselves worried about the potential dangers linked to artificial intelligence. Because he is convinced that artificial intelligence constitutes "our biggest existential threat", Elon Musk invested several million dollars in 2015 in 37 research projects designed to appraise its risks. For Nick Bostrom, the race for artificial intelligence is comparable to the race for atomic weapons during the cold war. Another example of fears linked to artificial intelligence is the project coordinated by Stanford University, which plans to monitor artificial intelligence for the next hundred years and to produce a report every five years to give an update on the state of advances in the sector. An open letter warning of these dangers gathered several supporters in 2015, including Nick Bostrom, Max Tegmark, Lord Martin Rees and Jaan Tallinn.[4]

Declarations of concern about the consequences of artificial intelligence have not prevented our prophets from imagining all sorts of revolutionary and extravagant projects. Elon Musk, for example, is involved with the Open AI initiative, an organisation founded in 2005, whose aims include developing artificial intelligence with a human face and extending the brain's capacity. Raymond Kurzweil estimates machines will surpass humans in 2045, at which moment man will at last have the chance to access augmented mental faculties. The researchers Marvin Minsky and Hans Moravec believe in the possibility of "mind uploading" or transferring a human mind to a machine. In this way, it would be possible to reconstitute the mind by simulating the brain's functioning: a kind of backup of our mind. In the same vein, Dmitry Itskov created Initiative 2045, to find out how to transfer individual consciousness to a computer, making life eternal. For Shimon Whiteson, professor at the University of Oxford, and finally for the American engineer Ben Goertzel, democracy will not be able to survive the robot revolution in the end. That researcher already foresees a scenario where we end up electing an artificial intelligence ourselves. But before we can back up and store the entirety of human intelligence on avatars, or other "nanorobots", we have to overcome the obstacle of the energy shortfall, as we mentioned earlier.

It is nevertheless true that the artificial intelligence market was due to go from 358 million dollars in 2016 to 31.2 billion by 2025 (Tratica 2018). Taking the 12 developed countries which together produce more than 50% of world economic production, the consultancy Accenture (Purdy and Daugherty 2016), working with Frontier Economics, estimates that artificial intelligence should enable growth rates to double between now and 2035. In the United States, the impact will be very significant, raising the annual growth rate from 2.6 to 4.6% in 2035 and in the UK from 2.5 to 3.9%. Even ageing Japan could potentially triple its growth rates. In France, artificial intelligence could enable a rise in productivity of around 20% and thus allow annual growth to develop from 1.7 to 2.9%. Economic forecasts may be optimistic about growth rates, but other studies put into sharp relief the dreadful impact on employment. Returning to the fear we have today of a kind of dehumanisation, we can legitimately cite the way in which Simone Weil defined the process of dehumanisation: "From its first property (the ability to turn a human being into a thing by the simple method of killing him) flows another, quite prodigious too in its own way, the ability to turn a human being into a thing while he is still alive. He is alive; he has a soul; and yet – he is a thing" (Weil 1941). Devastating words and an exceptional lucidity and intelligence reveal once again here how the unthinkable becomes reality; how those workers disappear in some way from history, turn into, or imitate, machines, with no discernible trace of rebellion anywhere. And with good reason. Anxiety about dismissal and unemployment is the most widely communicated feeling in those years of the Great Depression. Hence, this statement: "The body is sometimes exhausted in the evening, coming out of the factory, but the mind is exhausted all the time, and to a greater extent" (Weil 1941–1942). It is as if the philosopher experiences a kind of ethical dizziness in relation to this upside-down world. "The parts circulate with their dockets indicating their name, form and raw material: one could almost believe that it is they who are the people and the workers who are the interchangeable parts" (Weil 1941–1942). Obviously, artificial intelligence does not mean the death of the mind. Nevertheless, as many researchers believe, it is a question of a greater danger: of an uncontrolled future.

We can finally turn to Jean-Pierre Dupuy to close this description of a technological advance which is still more imagined than real. We can also call on Goethe and his famous poem The Sorcerer's

Apprentice. The former, a philosopher, author of a noteworthy work: *Pour un catastrophisme éclairé* (Dupuy 2004) (Enlightened Doomsaying) explains how much Ivan Illich made him see that humanity must be wary of three threats, not two. Alongside the force of nature and human violence, which humans have partly subdued, there is a third, which is much more difficult to fight because: "the enemy is ourselves. He has our own features, but we do not recognise him. Sometimes we diminish him along with nature; sometimes we turn him into a hateful and vengeful Nemesis" (Dupuy 2007). This evil actually comes from our capacity to act, "to set off irreversible, endless processes which can turn against us and take the form of hostile powers which destroy us", whereas beforehand, the sacred, religion or politics had the power to keep that capacity within certain limits, today human societies, guided by science and technology, are now capable of "setting off such processes within and on nature itself". Hannah Arendt wrote in The Human Condition that with modern technology, humans could claim to have "created" nature as they created history and thus had the power to "unchain natural processes of our own which would never have happened without us" by removing "the distinguishing boundaries which protected the world, the human artifice, from nature" (Arendt 1958).

## 3.2   Blockchain or the New Trust

Blockchain is the new Eldorado: total intermediation, the crazy idea that a direct relationship guarantees foolproof trust. This places the idea of that trust back at the heart of our considerations. Yet in the complex world of modernity or postmodernity, everyone thought risk was calculated and above all monitored and controlled. The crisis inflicted a cruel denial of this belief and in some ways betrayed the trust placed in the banking institutions and in their capacity as public powers for averting risk. Trust deserted households, businesses and even markets in state credit ratings. Yet this peak, which made people fear the worst, has precedents. During the 1990s, English-speaking political commentators and sociologists addressed the concept, showing that it was crucial for economic development and indeed for social and political development, but that it had a tendency to tail off, which necessitated measures to reconnect with democratic vitality and economic dynamism. Robert Putnam, for example, put the quality of human capital back at the heart of our society (Putnam 1995). In other words, he says "the quality of public life and the performance of social institutions" depends on the combination of standards, networks and trust. The gradual decline of this capital in the United States since the 1960s dangerously impoverishes American democratic life. Putnam believes this is the result of a multicultural society where "bonding capital", that is socialisation and trust between similar people, is not combined with bridging capital, that is socialisation and trust between people who do not resemble each other. As for Francis Fukuyama, he announced (Fukuyama 1995) a kind of triumph of trust as an inescapable factor in economic dynamism. Anthony Giddens tackled the same subject at around the same time. In *The Consequences of Modernity* (Giddens 1990), he showed that not only does modernity define itself by global trade nowadays, but that money, just like any symbolic guarantee, must be based on the absolute trust of the people in order to exist and function. For Giddens, in a modernity characterised by the evaluation of risks, trust is no longer placed in one person, but in the functioning of abstract entities, distant institutions and mechanisms in systems or networks which are the nature of "high modernity". If belief vanishes, the whole system stops functioning. Yet the globalisation of risk, its normalisation, impedes that high-risk trust. Dynamism and modernity, just like a runaway truck, raise anxieties and fears. Giddens did propose some well-founded methods, but you cannot re-establish trust by decree, except in authoritarian regimes. Faced with the absence of trust, the solution is to withdraw into a kind of pragmatic adjustment, a cynical pessimism, an enraged optimism or a no-less obstinate radicalisation. Thus, we know that powerful social inequalities are the breeding ground for a loss of trust by citizens towards the political and economic spheres and of the other in general. In short, they give rise to a kind of desperation about the future.

The promise of blockchain consists exactly of going from a centralised trust in institutions towards a decentralised one. From now on, trust depends on an IT network protocol capable of resolving the problem of the Byzantine Generals, which is a metaphor used in computing (Lamport et al. 1982). Resolving this problem amounts in fact to resolving the problem of trust. The problem is as follows: generals from the Byzantine army are camped around an enemy city and absolutely must establish a common battle plan to avoid defeat. But be careful, in among the generals there are men who could also turn out to be traitors and who will not hesitate to sow confusion through the group. How can we guarantee that the loyal generals can agree on the battle plan? How can we take account of suspect sources or information channels?

In a way, blockchain resolves this problem of communications breakdowns, whether they are accidental or malicious. It removes the need to go through a third party, that intermediary that traditionally casts a shadow over trust, over institutions and over exchanges between institutions. Its technology is based on a decentralised database which keeps a register, a sort of general accounts ledger, of activities and transactions completed, which is distributed via a peer-to-peer network. The register is public and it shows who holds what and who is carrying out which transaction. Its security is ensured by an encryption system. Transaction histories are bolted together in the form of linked blocks of data, secured with cryptography. This chain of blocks is the blockchain! The register, which is unalterable and forgery-proof, is then replicated on all the computers in the network.

Bettina Warburg compares blockchain to the online encyclopaedia Wikipedia. It is not an application or a business. On Wikipedia, everyone can read the articles freely. They are constantly updated and each change is tracked. "On Wikipedia, it's an open platform that stores words and images and the changes to that data over time. On the blockchain, you can think of it as an open infrastructure that stores many kinds of assets" (Warburg 2016). Where trust is concerned, blockchain is, in part, based on the same principles. With Wikipedia, we are dealing with a truth shared by entities which do not trust each other. For blockchain, all the nodes in the network do not need to know each other in order to carry out transactions. They are all capable of monitoring and verifying the chain themselves.

Don Tapscott considers blockchain to be the extension of the Internet. The Internet is the information tool. It enables us to copy and distribute information throughout the world, to send copies of our files, e-mails, photos, music, videos and more. But however revolutionary that may be, we are still dealing with copies. With blockchain, we are also exchanging with other people, but, most importantly, it makes it possible to exchange assets securely. We are not just dealing with copies any more, but with actual transactions! And the possibilities of exchanging assets are endless. It can involve money, bonds, title deeds, music, extracts from birth certificates, certificates of origin for food, ballot papers and so on. In this way, we will go from the Internet of information to the Internet of values and assets. Don Tapscott believes blockchain will also be a way for people to take back control of their personal data. The Internet may have encouraged the distribution of copies, but it has also allowed the emergence of a precious new value source: personal data.

In order to illustrate blockchain technology, we must discuss currency exchange, which is a perfect symbol of trust and always has been. The ancient world is marked by the transition from a barter economy towards using metal coins with interoperable units of account, consistent weights and identical shapes. Then, fiduciary currency makes its appearance: currency whose face value is not related to its physical form and composition. In order to make this a totally trust worthy currency, states simultaneously created the first national registers, such as the Napoleonic Land Register. Backed by institutions, for example states and international organisations, it became possible for an economy to be based on a coin or a banknote which has no intrinsic value. The monetary system has always been founded on the trust placed in these third-party institutions. Yet printing money is a specific feature reserved for states or international organisations. It was impossible to imagine a trusted currency which was not established as an institution, until an article appeared in 2008 signed by the mysterious Satoshi Nakamoto (2008), the creator of Bitcoin, the cryptocurrency. In his revolutionary paper, the author presented a computer protocol capable of resolving the Byzantine Generals Problem. His solution relied on a decentralised structure: blockchain. From now on, two agents who did not know each other could exchange assets without the need for the transaction to be certified by a third-party trusted authority. Nakamoto's identity is a mystery; there is a long list of potential authors, but it remains secret to this day.

Behind this iconic example, we find the three fundamental characteristics of blockchain technology. The first is disintermediation. Blockchain generates enough trust for users to exchange without supervision by a trusted third party. If we go back to the example of the banking system with Bitcoin, the transaction is validated by several nodes in the network called "miners". Their role is to check the validity of information transmitted by agents: the volume of funds available to the sender, the recipient and the volume to be transferred. To do this, the miners must solve a complex cryptographic problem which consumes computing power which can be jointly verified via a "proof of work". Once the miners agree on the validity of the proof of work, i.e. once there is a consensus, the transaction is integrated into a block of the chain. This consensus between the participants in the network renders obsolete the presence of an institution of reference, to guarantee the transaction.

The second characteristic of blockchain is the security guaranteed by two mechanisms: the decentralised framework and the cryptography method. Having a decentralised framework is a structural way of rendering the probability of the risk of data theft almost nil. The whole set of blocks in the blockchain is replicated in all the network nodes, not just inside one single server. This makes any pirating of the chain almost impossible. Another guarantee is the time stamp: each block is associated with a reference date and time aimed at proving the existence of that data before a certain date. In the blockchain, each new block's code is built on the block before it. The modification of one of the blocks would necessarily involve the modification of the entire set of blocks in the chain, which is impossible.

Finally, the last distinguishing trait of blockchain is its autonomy. For example, in the case of Bitcoin, tangible investment, computational power and storage space for the blockchain are guaranteed by the miners, the network nodes themselves. But their reward for this material support is paid in Bitcoin. This opportunity for financial gain then leads to competition between the miners to be the first to solve the cryptographic problem and present a proof of work. This gives them an incentive to invest in ever more powerful machines and guarantees the independence and speed of the Bitcoin protocol.

The future of blockchain is synonymous with the renewal of trust. Combined with the Internet of Things, blockchain makes the emergence of a new type of contract possible: smart contracts. This involves defining the terms of a contract in a blockchain and authorising its fulfilment when the conditions are checked and validated by a connected device. Let's take the more concrete example of the dispatch of a parcel and imagine that the data revealing its geolocation were the triggering factor of the contract. When the parcel is geolocated in the zone defined by the contract, the transaction is automatically carried out via the blockchain.

But we must take another step along this line of thought. What if blockchain was also the answer to keeping control over property and therefore guaranteeing it? One essential application of blockchain is actually to guarantee the longevity of land titles around the world, particularly in countries where there is a lot of political instability (Tapscott 2016). The Peruvian economist Hernando de Solo Polar believes land expropriation is the principle problem in terms of economic mobility. Having no guarantee of land titles is a source of uncertainty for numerous people across the world and exacerbates the constraints preventing them from building their future. By the same logic, blockchain is also the means of guaranteeing intellectual property. The arrival of the Internet has caused a lot of turmoil in the music industry. The protection of rights via a blockchain means people who would like to hear a piece could do so for free, or by paying a small sum, perhaps a few pennies, while for those who wish to use the content in another way, for example for commercial exploitation, the rate could be different and could go straight to the artist. In this way, artists, or any other creators of content, could become the owners of their own work again.

The promise of real disintermediation is also an opportunity to transform all intermediation platforms. The arrival of blockchain allows us to not only envisage decidedly lower charges, but most importantly to short-circuit all intermediation services. Take for example money transfer. Don Tapscott reminds us that it represents around 600 billion dollars per year. On average, such a transaction costs around 10 Euro cents and takes between four and seven days to complete. With a block-chain system, transferring money in one's native country in complete security will only take a few hours, for a much lower cost.

So everyone is imagining futures permeated by blockchain. In a recent study (World Economic Forum 2015), the World Economic Forum raises the possibility that governments could deduct taxes using a blockchain and that 10% of GDP could be stored on a block-chain. As a comparison, the total value of Bitcoin stored today on the blockchain is estimated at around 0.025% of world GDP in 2016. States are a bit infatuated with blockchain at the moment. The Bank of Canada announced via its senior deputy governor, Carolyn Wilkins, in June 2016 that it wanted to launch an experiment to study block-chain technology applied to currency. As for John Barrdear and Michael Kumhof (Barrdear and Kumhof 2016) from the Bank of England, a digital currency would, in their view, raise GDP by 3% because such a digital currency would lead to a reduction in interest rates, in tax rates and a lowering of transaction charges. They based this assess-ment on a barely credible hypothesis that 30% of UK GDP could be in digital currency form. Finally, in the United States, J. Christopher Giancarlo, Commissioner of the CFTC (Commodity Futures Trading Commission), considers that blockchain is the solution which will finally resolve the lack of visibility and "may finally give regulators trans-parency" (Giancarlo 2016). Because even after the initiatives and res-olutions for better regulation post-2008, "global regulators still do not have full visibility into the swaps trading portfolios of major financial institutions".

But it is the banks who are most affected. According to the consul-tancy McKinsey, international payment services yield 40% of the banks' annual sales revenue, which is around 1700 billion dollars per year. But despite technological progress, this sector has hardly developed since the beginning of the twentieth century. Andreas Adriano and Hunter Monroe (2016), two IMF economists, report on an assessment by the European Central Bank in 2012, according to which indirect costs, apart from commission, represent 1% of GDP and another, from the World

Bank, which estimates the cost of sending funds abroad at up to 8% of the sum transferred. One can imagine the upheaval that blockchain could create in that field of activity. We can see that blockchain technology can be considered a disruptive technology in the same way the car was to horse transport, or the Internet to the telephone. What this is really about is the principle of trust which it establishes on a new basis, which promises transparency and re-appropriation by individuals of their data.

As we know, thinking about trust is an issue which affects much more than banking. It cannot be summed up, in Locke's words as: "the bond of society" (Locke 1676). In some ways, we could say that trust has taken the place in the modern age that faith has in devout societies. It lacks that absolute, transcendent quality, but nevertheless it does need an overarching third party to exist. The mistake is to believe, like certain adherents of naïve liberalism, that a private contract between two people arises from a horizontal relationship, or a peer-to-peer one, as people love to boast in today's so-called disintermediated world. In fact, only the mediation of an outside entity of a larger order than the market sphere can avoid relationships based on calculations and self-interest. Get rid of religion and it will make a comeback, in the undoubtedly more unpredictable form of politics and institutions, but with values of common interest which go beyond the self-interested take of one or other party. Even the expression "trust in the currency" means a very different thing from the reciprocal trust between people.

So, what role does blockchain have in this story? The question of trust is not new, but it takes a new turn with the increasingly complex reality of our societies, as underlined by the sub-heading in the French edition of *Trust and Power* (Luhmann 2017) by sociologist Niklas Luhmann: "A mechanism for the reduction of social complexity".[5] This question has, in reality, been posed since the beginning of the modern age: since societies freed themselves from the theological-political model, which links trust with faith in God. The Moderns ask questions about trust as a means of reducing risk, or as the result of a rational calculation. It is a matter of formalising behaviours and institutionalising the contract of trust along the lines of forming an alliance, without which the future remains unforeseeable. This was very much Hume's message: for him, honouring a promise is an obligation. A promise thus invites the other party to trust the promise maker, but can also ruin the reputation of anyone who falls short of what they promise. Adam Smith also says that every economic transaction rests on interpersonal trust, a virtue which makes people willing to trade, while keeping an eye on their own interests at the same time. Here too, if trust is betrayed, the whole system is compromised. More than a century later, it is the turn of the German sociologist Georg Simmel to make trust "one of the most important synthesising forces in society"(Simmel 1908) which means one of the pillars of modern society is an attitude which eludes pure rationality. Trust is a gamble: it does not claim to know everything. It is fragile and can be betrayed. Simmel goes even further when discussing currency: "it contains a further element of social-psychological, quasi-religious faith. The feeling of personal security that the possession of money gives is perhaps the most concentrated and pointed form and manifestation of trust in the socio-political organization and order" (Simmel 1900).

We can see that the whole person, not someone with their emotions cut out, is drawn towards trust, whether it is interpersonal, or an abstract principle based on prescriptive ideas, supported by institutions which must themselves be reliable.

There is a dark side, a black version, which people try to conceal in any way they can. This is despite the fact that fear has tormented us since the financial crisis of 2008, and the 1929 crisis before it, when panic ruled to such an extent that people believed that it was the end of the world. Despite the quite widespread climate of defiance in which France and other Western societies live, it is useful to remember that a third party is watching over us: politics, at its most noble, is one of the most effective barriers against the threat of a contagious fear and mistrust which could sign the death warrant of the financial system. No blockchain could take its place.

## 3.3    Big Data or the Disappearance of Free Will

The disappearance of free will, or the appearance of servitude, refers in the very first instance to total dependency, in the form of slavery or serfdom. Slavery and serfdom may date back into history, but they are not accidents of history. They were an integral part of the economic activity of the ancient city, of feudal society and of the different regions of the world which organised the slave trade. The precondition which unites all those situations is that servitude is nothing other than the state of total dependence. It can apply to a person, a class of individuals or indeed an entire country, which is conquered and entirely subservient to the authority, control and whim of another entity, whether that is a person or a group. Indispensable to the economy, but trapped by the implacable logic of power, serfs and slaves see their status as human beings stripped away, to a greater or lesser extent. The great democrat Aristotle believed that those who do not belong to the city state, who are not citizens, have nothing human about them. "The [city] state belongs to the class of objects which exist by nature and […] man is by nature a political animal. Anyone who by his nature and simply by ill-luck has no state is either too bad or too good, either subhuman or superhuman – he is like the war-mad man condemned in Homer's words as 'having no family, no law, no home'; for he, and who is such by nature, is mad on war: he is a non-co-operator, like an isolated piece in a game of draughts" (Aristotle 1981). He adds: "Therefore whenever there is the same wide discrepancy between human beings as there is between soul and body or between man and beast, then those whose condition is such that their function is the use of their bodies and nothing better can be expected of them, those, I say, are slaves by nature". Conversely, Plato's ideal city did not harbour a single slave. Servitude, for him, has a different nature: it is at the heart of the city state when the democratic government, which is the most disordered type of government, inevitably gives way to tyranny. He explains: "The only likely reaction to excessive freedom, whether for an individual or for a city, is excessive slavery" (Plato 2000). In the hierarchy of governments, the last, tyranny, is the worst enemy of freedom and signals the wrecking of humanity. The tyrant begins by pandering to the crowd's desires: by giving them goods and riches. Then, he governs for himself, surrounding himself with a guard loyal only to him. Then, to divert the hatred of his fellow citizens, finds the distraction of choice: war. "Isn't a tyrant always bound to be stirring up war?" (Plato 2000).

Let's pass through the centuries: past the Late Roman Empire, with its one million inhabitants, 1% nobles and one-third slaves. Let's turn to two figures who are allies in humanism: Erasmus and More. They were not only contemporaries of the wars of religion, but also of land enclosure in England and its ravages on society: suddenly depriving peasants of any means of subsistence. Removed from their context, these words which Erasmus addressed to More are completely apt here: "I find death sweeter than slavery". Those two great figures of emerging humanism paid a high price for their freedom.

When we mention Big Data, Winston Smith in George Orwell's *Nineteen Eighty-Four* inevitably comes to mind. It is an easy point to make, but "Big Brother is watching you", even with the more acceptable face of service technology, is not so far from today's reality… The exponential growth of Big Data is incomprehensible to most people. Those who are better informed see it as a source of promise or alarm.

Let's return to what Big Data is. It is defined as the combination of the technology, infrastructure and services which enable the collection, storage and analysis of data. And in this sector of technology, data is the raw material. The mathematician Clive Humby was one of the first, in 2006, to draw a parallel between the role of petrol in the second industrial revolution and that of data in the digital revolution. As with petrol, in order to exploit its full potential, it is first necessary to extract the data in its crude form, then "refine" it. The data boom, boosted by the emergence of the Internet of Things, actually contributes to the sustainability of Big Data. Moreover, the Internet of Things could help double the size of the digital world every two years. Some commentators (EMC and IDC 2014) say the size of the digital world could represent 44,000 gigabytes by 2020, which is around 10 times more than in 2013. From now on, every connected device becomes a new, independent, functioning system with its own infrastructure. This enables the exponential explosion in available data. Estimates for the number of connected devices by 2020 vary between 30 billion and 220 billion, depending on the source used (Institut G9+ 2013); nevertheless, they all confirm that the digitisation of the real world is gradually becoming omnipresent in our environment.

If we want to understand Big Data, the classic definition involves the three Vs (Institut Montaigne 2015): volume, variety and velocity.

"Volume", firstly, indicates that we are dealing with a very significant volume of data. In 2011, we exceeded a zettabyte[6] for the first time. In one year, 200 times more data was collected and recorded than everything collected and recorded up to that point. Viktor Mayer-Schönberger and Kenneth Cukier have attempted to give a clearer representation of this using a comparison. They said that if we explained the volume in the form of printed books, it would be equivalent to covering the entire surface of the United States with 52 layers of them. If you explain it using piled-up CD-ROMs, the volume would correspond to five piles which would reach the moon. We passed the level of 4.4 zettabytes in 2013 and it is estimated that we will reach a volume of 44 zettabytes in 2020 (EMC and IDC 2014)!

The second characteristic of Big Data lies in the "variety" of exploitable data. Unlike former techniques, where it was absolutely necessary to use structured data, Big Data can now exploit all types of non-structured data available. This means that all types of data are exploitable: text, images, multimedia content, digital footprints, connected devices' data and so on.

Finally, the last characteristic: "Velocity". This refers as much to the speed of the information as to the speed of the simultaneous processing of data. From now on, for better or for worse, data is analysed in real time. For example, in Canada, it has been possible to detect infections in premature babies 24 hours before the first symptoms appeared, by exploiting information in real time. In the case of driverless cars, communications between them and connected infrastructure actually avoid accidents.

But in this environment ruled by data, the risk of being under surveillance is a particular source of fear. According to Bruce Schneier, a specialist in computer security, surveillance has become the business model which the Internet relies on. He even describes it as the "golden age of surveillance". One aspect of this surveillance is the concentration of enormous databases of private information, and even the collection of personal data via certain state programmes, such as that run by the NSA.[7] Another aspect is the threat of Internet piracy, which also affects states and major companies. We had to wait three years to learn that more than a billion accounts were hacked in 2013. Once the data has been stolen, it is often put up for sale on the Dark Web.

Carlo Ratti and Dirk Helbing use the "price of anarchy" to highlight very clearly another fear linked to Big Data (Ratti and Helbing 2016). In game theory, the "price of anarchy" represents the loss of effectiveness due to the individual actions of each agent motivated by their own interests, rather than the interests of the system they are part of. The classic example to illustrate this concept is the management of urban traffic. A centralised, hierarchical approach to managing traffic in cities is favoured, over one which would let each driver make their own choices in the hope that the combination of individual drivers' choices might produce an acceptable arrangement. We could therefore say that the centralised approach is the one which reduces the cost of anarchy. If we draw a parallel with the boom in available data, we can see that the power of the predictive models used in Big Data could be the right solution for reducing the price of anarchy. But if we accept such a model, would it amount to allowing the end of chance and free will, and choosing a new servitude?

Let's take the example of large companies' recommendations for purchases of goods online. Based on the behaviour of consumers and their buying history, they suggest to users the next books or items they should order. Using the past, they propose to predict the future and at the same time reduce the cost of anarchy. But is the next book which meets the reader's needs necessarily related to the previous ten books they bought? Or should we consider that the ideal book might be the very one which provokes a new, unpredictable, unexpected feeling in the reader? How can you deeply move a reader if the books they are offered are calibrated using a historic, predictive model? By relying too much on algorithms to optimise the selling process, do we not risk suffocating everyone's unpredictability and creativity?

Slavery is one of the oldest and most frequently tackled topics for philosophers from all eras, who come at it with or without an emphasis on determinism, whether collective or individual. It would be laborious, indeed impossible to summarise it here without rattling off platitudes. But let's return to the ancient Greeks, since the purpose of the *polis* was to ensure the freedom of citizens (and not slaves). This freedom was understood as a "power to begin", that is to break from the course of events, to initiate an action which demonstrates the individual is just that: different from all the others. Yet, according to Arendt, the modern age's blessing of the "*laborans*" animal, the individual hooked up to machines and no longer communicating with others, along with the withering away of public space and the surge in growth of private space, run counter to the purpose of *polis,* or politics in the Greek sense. Arendt adds that standardisation has stripped the term "liberty" of its meaning and that technically, the purpose of the *polis* today is to keep humans alive. Take away any discussion with others, and we are left with total alienation from the world: a statement which might be shared by André Gorz, with his frantic defence of the autonomy of the individual.

However Arendt's analysis does nevertheless raise questions, in the convergence she finds between totalitarian regimes and modern times over her definition of freedom. Both models sit comfortably with the servitude of most of the population: in the name of different ideologies, but for many, equally all-consuming reasons. We find echoes here of some of the texts on technical progress and the subjugation of people cited earlier. Right now, Big Data is capable of taking on a "watch and punish" type of thinking, which is the brutal logic of all authoritarian powers: the logic of control and domination of the masses, often at work in human history.

Let's remember that the promise of digital services is to make our daily lives easier. We now have watches which monitor our health or our heart rate overnight, or use a geolocation service which tells us the ideal time to set off on a journey. In the end, why not accept the fact that a private company can keep so much personal data, that it knows what we look at on the Internet, or that it can even collect medical information? We communicated personal information to third parties before the arrival of the Internet, whether it was to bankers, doctors or tax authorities. However, the difference is that until this point, the information was transmitted erratically, whereas now, one single company has the ability to centralise all sorts of private medical or financial information, linked to areas of interest, consumption behaviour, political orientation, geolocation, etc. If we entrust all of that data to the same organisation so that it can anticipate our desires for us, or promote products which are tailor-made for our profiles, we are obviously risking a new kind of servitude.

According to a recent report, the worldwide exploitation of open data could generate a potential market of 3000 billion dollars (McKinsey 2013) for the seven industries studied. What is more, its analysis of 150 case studies concluded that the Internet of Things could have a potential economic impact of between 3900 and 11,100 billions of dollars by 2025. At its peak, the Internet of Things could represent 11% of the global economy. At the French national level now, the consultancy A.T. Kearney (Institut Montaigne 2015) values the market in connected devices at 15 billion Euros by 2020 and 23 billion by 2025. It values the potential for creating value from the Internet of Things at between 74 billion Euros in 2020 and 138 billion in 2025. The economic stakes are very high. We can see that stopping this movement will be very difficult.

Behind this very predictable difficulty, our firm purpose in concluding our reflection on Big Data is to return to La Boétie's famous reflection on voluntary servitude. Five centuries later, it still torments us, as it haunted Hannah Arendt during the Eichmann trial in Jerusalem in April 1961. How can we conceive that people could, of their own accord, abandon their dignity and renounce thought, and that they didn't draw lessons from that experience? "It is therefore the inhabitants themselves who permit, or, rather, bring about, their own subjection, since by ceasing to submit they would put an end to their servitude. A people enslaves itself, cuts its own throat, when having a choice between being vassals and being free men, it deserts its liberties and takes on the yoke, gives consent to its own misery, or rather, apparently welcomes it" (De La Boétie 1576). La Boétie tries to elicit a response to this lack of reason: humanity's tendency to turn its back on what it cherishes most and accept submission and indifference. But we have already discussed this, in the context of dehumanisation.

## 3.4    Genetic Engineering or the Man-God

According to the Bible, pride is a basic human inclination. Adam's descendants, united by the same language, pit their collective power against God's when they build the tower of Babel. Somehow, out of a kind of imitative desire, they attempt to compete with the divine power: an initiative doomed to failure. God punishes them by scattering them and creating multiple languages. "Behold, they are one people and they have all one language; and this is what they begin to do; and now nothing will be withholden from them which they purpose to do" (ed. Herz 1960). One could think of this punishment as a blessing, in that it establishes the diversity of human beings. However, the truth remains that whoever gets to the top of the tower is tempted to think of themselves as God. In the same way, the Gods of Olympus dispatch Nemesis to punish human excess: hubris, denounced by the philosophers of Ancient Greece. Yet, although our modern societies no longer live under God's influence and have eradicated all traces of the divine, pride often remains unchecked, with all the dangers it entails. Advances in science and technology enable us to compete with the prophets of the apocalypse and bring about the end of the world without any "chosen people", that is, the destruction of the Earth's ecosystem. Somehow we have granted ourselves a power which the ancients kept only for the divine. However, as Bertrand Vergely says in *La Tentation de l'homme-Dieu* (The Temptation of the Man-God) (Vergely 2015), this particular fantasy has plenty of life left in it. For the prophets of post-humanism, it is a matter of breaching that final frontier which differentiates humans from God: death. This man-god, Vergely says, who already has the power to destroy life, has decided his death belongs to him and that he can choose its date and time through euthanasia. But what is more, he has set out to put an end to death, or "to put death to death" as Laurent Alexandre (2011) says. Whether it be the transhumanist project to drive away and defeat disease, or an Italian neurosurgeon's crazy project to transplant the head of one person on the body of another, all of these sometimes monstrous initiatives demonstrate not only the desire to get rid of death, but also the desire to get rid of humans, of history and of time. Or, to put it in a more starkly, to get rid of life. The temptation of the man-god is a naïve fantasy. It is the farthest it is possible to get from that humanist tradition which gives humans all their dignity: a life worth living.

Biology has been at the heart of debate for several decades. These debates do not focus on its major advancements as much as on the tempting possibility it has to entirely dominate other disciplines and for the ethical questions raised by applications of its discoveries. We can illustrate the development of medical technological advances via three examples of discoveries linked to genetic engineering: the CRISPR-Cas9 method of genetic manipulation, personalised medicine based on the study of the human genome and finally, 3D bioprinting.

The CRISPR-Cas9 genetic engineering technique was declared the discovery of 2015 and then, almost simultaneously, associated with a weapon of mass destruction by the US National Security Agency (NSA 2016) in 2016. This new technique shook up the scientific community because it offers the opportunity to engineer the DNA of living beings with disconcerting ease. Thanks to this technique, replacing one gene with another or modifying DNA has become a quick procedure which is easy to implement. Above all, it has made this type of operation very cheap. Scientists can, at last, make changes in the DNA of cells and attempt to treat genetic illnesses!

CRISPR-Cas9 is in fact an enzyme, capable of detecting a specific part of DNA and destroying it. CRISPRs, or "Clustered Regularly Interspaced Short Palindromic Repeats", are families of repeated DNA sequences. From 1987 onwards, Japanese researchers had noticed unknown sequences in certain bacteria, which were present between certain other basic sequences that make up DNA. In 2005, it was finally discovered that these "unknown" sequences actually corresponded to the sequences of bacteriophages: viruses which only attack bacteria. When a bacterium is attacked by a virus, it retains the virus DNA in its memory in order to be able to fight it if it returns. It is a type of immune system. When the bacterium detects a virus, it will create an RNA (ribonucleic acid) which corresponds to the virus, but also an enzyme, Cas9. Cas9 will then integrate with the RNA and attach itself to the virus to cut its DNA and destroy it.

Researchers inspired by this mechanism devised a technique in 2012 which enabled them to synthesise the Cas9 enzyme. By cutting a conventional cell, it becomes possible to introduce a DNA mutation into it: a piece of modified DNA which the cell's enzymes will pick up. In fact, such DNA "scissors" existed already, but the technique had never been so simple to carry out. Jennifer Doudna, one of the researchers behind this method with Emmanuelle Charpentier and Feng Zhang,[8] sums up the method as follows. For her, CRISPR-Cas9 is a tool which enables the precise, easy and quick modification of genes. There is a protein which acts as scissors and which cuts the DNA sequence and a molecule of RNA which leads the scissors to the desired location on the genome. It is a bit like editing, but for genes. We can remove a gene, add one and even modify one single letter of a gene. And we can do all this for any species!

Cutting out and replacing defective genes has enabled many experiments, including one concerning the vision of rats suffering from blindness due to an inherited illness. This technique seems to offer fundamental research the means to improve our understanding of the brain and the role certain genes have on it. In therapeutics, this method makes it possible to eliminate, correct or replace the gene responsible for a mental illness. The cost of the technique is so low that researchers can pursue extremely varied studies. In theory, it has become possible to correct hereditary genetic diseases by modifying germ cells. In other words, humans are now capable of modifying the human species!

The CRISPR-Cas9 method has also led to the appearance of the "gene drive", that is the transmission of a precise gene: the propagation of a mutation at the level of an entire population. By modifying a portion of the DNA of a few people within a given population, it is possible, within a few generations, to contaminate an entire population with the "gene drive": the suite of genes modified by humans. More precisely, if ten genetically modified individuals carrying a "gene drive" suite of genes are introduced into a given population of 100,000 people, after 12–15 generations, 99% of the individuals (Morizot and Orgo 2016) will be carriers of the gene drive suite. However, conversely, a simple genetic mutation present in the same proportion of the population would have disappeared from the population after just a few generations. This represents a new phase in genetic engineering which henceforth gives humans the power to transform species. We can, of course, underline the immense progress this entails. Notably, this method could prevent the propagation of certain invasive species, malaria, dengue fever, chikungunya virus, yellow fever or even Asian carp in the Great Lakes…It could also block the progress of new virus pathogens like the

Zika virus, often linked to underdeveloped cranial vault in newborns in Latin America. But such power, wielded over humans and ecosystems by these examples of genetic engineering and proliferation, is quite simply terrifying.

Jennifer Doudna, one of the technique's inventors, does not hide her concern about the possible results produced by some of its practitioners. Remember, this method could eventually enable the conception of humans with improved characteristics, for example stronger bones, or new qualities which make them more attractive, or choose their eye colour, etc. What is frightening is that the technique is so simple that Jennifer Doudna considers it to be within almost everyone's reach. It could very well be used for criminal ends, for example, if someone wanted to genetically transform bacteria or neutral viruses into pathogenic micro-organisms capable of secreting deadly poisons. Doudna estimates that someone with basic qualifications and easily accessible equipment would be capable of carrying out the genetic engineering technique.

But we are not just talking about fictional cases. There have already been several risky experiments. For example, in April 2015, the team of Junjiu Huang at the University of Sun-Yat-Sen in Canton tried to treat human embryos which carried an abnormal gene which causes a potentially life-threatening blood disorder, beta-thalassaemia, in an experiment with the CRISPR-Cas9 method. Once again, during spring 2016, a new Chinese university used the CRISPR-Cas9 method on human embryos, except that this time, it decided to apply it to living humans. Lu You, an oncologist at the University of Sichuan in China, selected ten patients, all suffering from incurable lung cancer. His aim is to strengthen the patients' bodies' natural defences so that they can fight the tumour. We could also cite other, somewhat eccentric projects, such as that of Harvard professor of genetics George Church, who wants to bring back the woolly mammoth, which disappeared around 4000 years ago, by recovering its genome and implanting it into an Asian elephant. He says it will take between seven to ten years to reach this goal.

The second innovation linked to genetics is personalised medicine, which is based on the study of the genome. The genome represents the entire DNA of a living being. It contains genetic information, like a sort of encyclopaedia of an individual or a species. These days, sequencing an entire human genome costs less than 1000 dollars. It is worth remembering that the first sequencing of a human genome, published in 2003, took more than ten years and cost three billion dollars! This advance makes so-called personalised medicine possible today. This is a branch of medicine which treats each patient individually according to their own genetic characteristics.

High-speed sequencing of the human genome could therefore lead to a better understanding of patients, a better diagnosis of their problems and an assessment of their risks of contracting certain illnesses, such as diabetes, Alzheimer's disease or cancer. It could also make it possible eventually to measure patients' risks of transmitting potential genetic disorders to their children. It could also simply detect a potential allergy to a medicine. There are already many projects which are currently making advanced use of this new knowledge: Genomic England, in the UK, envisages sequencing 100,000 genomes of British people, most of whom suffer from hereditary illnesses or cancers. The Beijing Genomics Institute is attempting to index the genes of exceptionally gifted people with IQs of more than 160. Sequencing can even be carried out on foetuses, because some of their DNA circulates in the mother's blood. This approach avoids the need for amniocentesis and could allow for quick detection of certain conditions and genetic mutations. It has been estimated (McKinsey 2013) that genomic science could have a potential economic impact of between 700 billion and 1600 billion dollars by 2025. Moreover, the impact of the prevention of illnesses and the applications for treatment has itself been costed at between 500 and 1200 billion dollars.

And finally we come to 3D bioprinting. A point of convergence between physics, biology, mechanics and information technology, bioprinting makes it possible to create bespoke tissues and organs from the patient's cells, minimising the risk of rejection. It is now possible to reproduce sections of bone, muscle, cartilage and skin using a 3D printer. For example, the 3D printer is capable of reconstructing skin, layer by layer, from an image modelled virtually by a computer. Models of tissue engineering already existed, but the innovation of bioprinting is the speed and finesse with which it is carried out. The speed of printing equates to less than two minutes per square centimetre. This means it would be possible to reconstitute an entire face in five hours.

Practical applications of this may be limited at the moment, but we can anticipate a great number of possibilities arising from such a technique. The main aim would be to use it as a priority for biologically printed organ transplants, skin and prostheses, thus avoiding the risk of rejection and lowering the costs of medical care. Such technology could also be used to print individualised tissues for experimental use. For example, in cancer patients, reconstructing tissues in 3D could make it possible to anticipate the patient's reaction, whether positive or negative, to different types of chemotherapy. In the same way, routinely printing diseased tissues such as cancerous tumours would enable scientists to test their composition and target the most effective molecules for an individual patient.

These three major technological breakthroughs: the CRISPR-Cas9 method, personalised medicine via genome studies and 3D bioprinting, do confirm that behind everything, prolonging human life remains the priority of our "prophets". For example, Peter Thiel, the famous founder of PayPal, is also one of the man investors in the Methuselah Foundation, named after the biblical character who lived for 969 years. The foundation worked on the production of medicines to limit the impact of ageing on cells.

We must give the last word, the ethical one, to Henri Atlan, eminent biologist and philosopher (Albaret and Padis 2006)[9]: "In fact, ethical consideration should only apply to questions with concrete applications and well-defined techniques and not to research programmes which are still a long way off from such applications". He continues by suggesting "the stage at which the brakes should be applied: never abstain from growing in knowledge and wisdom, right up to the point of being capable of making a perfect artificial human, to then perfect *imitatio dei*, the highest vocation of humanity; but abstain from making it once able to do so. There is no ethical reason to stop or slow down the research into developing these applications, if only because the applications cannot in practice ever be anticipated – nor obviously can their beneficial or harmful effects – far in advance of their actual creation".

As Henri Atlan expresses so well, and as we indicated at the beginning of this chapter, the desire to be all-powerful, to compete with God, is a recurring theme in human history. What is more surprising is the unanimity among our prophets of the "augmented human" (or post-human, as it has long been called by Dominique Lecourt). Thiel, Ellison, Brin, Itskov, Kurzweil, the late Minsky, Diamantis, Ishiguro, Dubrovsky and many others have already put forward arguments "to ensure the survival of civilisation, build a bright future for all mankind, reach new goals and create new meanings and values for a human, ethical and high-tech future" (GF 2045 2013). The least one can say about this is that it does not err on the side of restraint. This "sin" of desiring omnipotence is well known. The Greeks called it *hubris* and were horrified by it. It is a synonym for lack of restraint, reprehensible pride and violent passion. It amounted to a crime and its author was punished accordingly. And that only concerned the reckless madness of competing with immortal gods. What if we discovered a technological tool capable of making us, or rather some of us, all-powerful: to allow us to be on first-name terms with death? The Berlin Jewish historian and philosopher Gershom Scholem, a great expert in Kabbalah, was invited by the Weizmann Institute in Rehovot, Israel, to give an address to mark the construction of the first computer in Israel in 1965. He let it be known that he would give it the name Golem 1 (Scholem 1991). In the light of history, he was certainly not wrong.

## Notes

1. The Golem (Hebrew word for "embryo", "shapeless" or "unfinished") is an artificial being from Jewish mysticism and legend, usually humanoid, made of clay, incapable of speech and deprived of free will, made in order to assist or defend its creator
2. Narrow Artificial Intelligence
3. AGI: Artificial General Intelligence
4. Future of Life Institute (2014) *Research Priorities for Robust and Beneficial Artificial Intelligence*. Available via http://futureoflife.org/ai-open-letter/

5. Luhmann, N. (2006). *La Confiance – Un mécanisme de reduction de la complexité sociale*. Paris: Economica
6. 1 zettabyte $= 10^{21}$ bytes.
7. Such as the PRISM surveillance programme revealed by Edward Snowden in 2013
8. Feng Zhang disputes the precedence of Jennifer Doudna and Emmanuelle Charpentier's discovery
9. Albaret I & Padis MO (2006) *Interview with Henri Atlan "La recherche scientifique face à la réflexion éthique"* [Scientific Research Versus Ethical Thinking]. Esprit (Supplement) Paris. Quote here translated by D. Leifer. Available via https://esprit.presse.fr/actual-ites/marc-olivier-padis-et-propos-recueillis-par-isabelle-albaret/la-recherche-scientifique-face-a-la-reflexion-ethique-39894.

# References

Adriano, A., & Monroe, H. (2016). *The Internet of Trust, in Finance & Development*. Available via https://www.imf.org/external/pubs/ft/fandd/2016/06/adriano.htm.

Albaret, I., & Padis, M. O. (2006). *Interview with Henri Atlan "La recherche scientifique face à la réflexion éthique"* [Scientific Research Versus Ethical Thinking]. Esprit (Supplement) Paris. Quote here translated by D. Leifer. Available via https://esprit.presse.fr/actual-ites/marc-olivier-padis-et-propos-recueillis-par-isabelle-albaret/la-recherche-scientifique-face-a-la-reflexion-ethique-39894.

Alexandre, L. (2011). *La Mort de la mort – Comment la technomédicine va bouleverser l'humanité* [The Death of Death—How Techno-Medicine Will Disrupt Humanity]. Paris: Jean-Claude Lattès.

Arendt, H. (1958). *The Human Condition* (2nd ed.). Intro Margaret Canovan (1998). Chicago: University of Chicago Press.

Aristotle. (1981). *The Politics*, 1, II. (T. A. Sinclair, Trans.). Revised T. J. Saunders. London: Penguin.

Barrdear, J., & Kumhof, M. (2016). *The Macroeconomics of Central Bank Issued Digital Currencies* (Bank of England Staff Working Paper No. 605). Available at https://www.bankofengland.co.uk/-/media/boe/files/working-pa-per/2016/the-macroeconomics-of-central-bank-issued-digital-currencies.pdf?la=en&hash=341B602838707E5D6FC26884588C912A721B1DC1.

De La Boétie, E. (1576). *The Discourse of Voluntary Servitude* (H. Kurz, Trans., 1975). Auburn, AL: The Mises Institute.

Dupuy, J. P. (2004). *Pour un catastrophisme éclairé, Seuil, coll.* Paris: Points Essais.

Dupuy, J. P. (2007). *D'Ivan Illich aux nanotechnologies. Prevenir la catastrophe?* (Interview). Paris: Esprit. Available via https://esprit.presse.fr/article/jean-pierre-dupuy/d-ivan-illich-aux-nanotechnologies-prevenir-la-catastrophe-entretien-13958.

EMC and IDC. (2014). *The Digital Universe of Opportunities: Rich Data & the Increasing Value of the Internet of Things.* Available via https://www.emc.com/collateral/analyst-reports/idc-digital-universe-2014.pdf.

Fukuyama, F. (1995). *Trust: Social Virtues and the Creation of Prosperity.* London: Penguin.

Genesis, Chapter 11, Verse 6. In C. H. Hertz (Ed.), *The Pentateuch and Haftarahs* (2nd ed.) (1960). London: Soncino Press.

Giancarlo, C. J. (2016). *Keynote Address of CFTC Before the Cato Institute: Cryptocurrency: The Policy Challenges of a Decentralized Revolution.* Available via https://www.cftc.gov/PressRoom/SpeechesTestimony/opagiancarlo-14.

Giddens, A. (1990). *The Consequences of Modernity.* Oxford: Polity Press.

Global Future 2045. (2013). *Mission.* Available via http://gf2045.com/.

Institut G9+. (2013). *Les nouveaux eldorados de l'économie connectée* [The New Eldorados of the Connected Economy]. Available via https://www.g9plus.org/publications/3-livre-blanc-les-nouveaux-eldorados-de-l-economie-connectee.

Institut Montaigne. (2015). *Big Data and the Internet of Things: Making France a Leader in the Digital Revolution.* Available via https://www.institutmontaigne.org/en/publications/big-data-and-internet-things.

Lamport, L., Shostak, R., & Pease, M. (1982). The Byzantine Generals Problem. *ACM Transactions on Programming Languages and Systems, 4*(3), 382–401.

Locke, J, (1676). *Essays on the Law of Nature.* Oxford: Oxford University Press (2002).

Luhmann, N. (2017). *Trust and Power* (H. Davis, J. Raffan, & K. Rooney, Trans.). Cambridge, UK: Polity Press.

McKinsey Global Institute. (2013). *Disruptive Technologies: Advances That Will Transform Life, Business and the Global Economy.* Available via https://www.mckinsey.com/business-functions/digital-mckinsey/our-insights/disruptive-technologies.

McKinsey Global Institute, McKinsey Center for Government, McKinsey Business Technology Office. (2013). *Open Data: Unlocking Innovation and Performance with Liquid Information.* Available via https://www.mckinsey.com/~/media/McKinsey/Business%20Functions/McKinsey%20Digital/Our%20Insights/Open%20data%20Unlocking%20innovation%20and%20performance%20with%20liquid%20information/MGI_Open_data_FullReport_Oct2013.ashx.

Minsky, M., & Papert, S. A. (1969). *Perceptrons.* Cambridge, MA, USA: MIT Press (1987).

Morizot, B., & Orgo, V. (2016). *Faut-il relâcher le "gene drive" dans la nature? Enjeux civilisationnels des "OGMs sauvages"* [Must We Release the Gene Drive into the Wild? The Implications of "Wild GMOs" for Civilisation]. Available at http://www.normalesup.org/~vorgogoz/gene-drive.html.

Müller, V. C., & Bostrom, N. (2016). Future Progress in Artificial Intelligence: A Survey of Expert Opinion. In *Fundamental Issues of Artificial Intelligence* (pp. 553–571). Berlin: Springer.

Nakamoto, S. (2008). *Bitcoin: A Peer-to-Peer Electronic Cash System.* Available via https://bitcoin.org/en/bitcoin-paper.

National Security Agency. (2016). *Worldwide Threat Assessment of the US Intelligence Community, Senate Armed Services Committee.* NAS Report. Available via https://www.dni.gov/files/documents/Newsroom/Testimonies/SSCI%20Unclassified%20SFR%20-%20Final.pdf.

Plato. (2000). *The Republic* (T. Griffith, Trans.). Cambridge: Cambridge University Press.

Purdy, M., & Daugherty, P. (2016). *Why Artificial Intelligence Is the Future of Growth, Accenture.* Available via https://www.accenture.com/t20170524T055435__w__/ca-en/_acnmedia/PDF-52/Accenture-Why-AI-is-the-Future-of-Growth.pdf.

Putnam, R. (1995). Bowling Alone: America's Declining Social Capital. *Journal of Democracy, 6*(1), 65–78.

Ratti, C., & Helbing, D. (2016). *The Hidden Danger of Big Data.* Project Syndicate. Available via https://www.project-syndicate.org/commentary/data-optimization-danger-by-carlo-ratti-and-dirk-helbing-2016-08?barrier=accesspaylog.

Scholem, G. (1991). *The Messianic Idea in Judaism and Other Essays on Jewish Spirituality.* New York: Schocken Books.

Simmel, G. (1900). *The Philosophy of Money* (T. Bottomore & D. Frisby, Trans., 2004). London: Routledge.

Simmel, G. (1908). *Sociologie. Études sur les forms de la socialisation*. Paris: Presses Universitaires de France (2013).

Tapscott, D. (2016). *How the Blockchain Is Changing Money and Business*. TED Summit. Available via https://www.ted.com/talks/don_tapscott_how_the_blockchain_is_changing_money_and_business?language=en.

Tratica. (2018). *Artificial Intelligence for Enterprise Applications*. Available via https://www.tractica.com/research/artificial-intelligence-for-enterprise-applications/.

Turing, A. M. (1950, October 1). Computing Machinery and Intelligence. *Mind, LIX*(236), 433–460. Available via https://doi.org/10.1093/mind/LIX.236.433.

Vergely, B. (2015). *La Tentation de l'homme dieu* [The Temptation of the Man-God]. Paris: Le Passeur.

Warburg, B. (2016). *How the Blockchain Will Radically Transform the Economy*. TED Summit. Available via https://www.ted.com/talks/bettina_warburg_how_the_blockchain_will_radically_transform_the_economy?language=en.

Weil, S. (1941). The Iliad, or the Poem of Force (M. McCarthy, Trans.). In *War and the Iliad* (2005). New York: The New York Review of Books.

Weil, S. (1941–1942). Expérience de la vie d'usine [Experience of Factory Life]. In *Oeuvres completes*, tome II, vol 2, Paris: Gallimard.

Wiener, N. (1948). *Cybernetics, or Communication and Control in the Animal and the Machine*. Cambridge, USA: The Massachusetts Institute of Technology.

Wiener, N. (1964). *God & Golem Inc: A Comment on Certain Points Where Cybernetics Impinges on Region*. Cambridge, MA: MIT.

World Economic Forum. (2015). *Deep Shift, Technology Tipping Points and Societal Impact*. Survey Report. Available via http://www3.weforum.org/docs/WEF_GAC15_Technological_Tipping_Points_report_2015.pdf.

# 4

# A Shattered Labour Market

The relationship between technological progress and employment has been a classic issue of economics debate for more than two centuries. But while it might appear to be a rather academic question for some, its real stakeholders experience it as a tragedy. In fact, once innovation appears, fear quickly follows: as experienced by the Luddites, or the Lyon Canut silk workers, but also well before that, by Greek soldiers confronted with catapults. There are certain thinkers who are a bit short on ideas and envisage, like Rifkin, the end of work. They believe even a drastic reduction in working hours would not allow us to rebalance labour supply and demand. Others are overly optimistic, like Sauvy, and believe new employment will be created almost simultaneously as jobs disappear due to the so-called technological unemployment where machines replace people. This is an eternal debate with no clear-cut answers, but with some obvious facts: in the short term, more jobs have been destroyed than created, and there is no certainty at all about what will happen in the future. Today's thinkers limit themselves to discussing the digital revolution, which represents only one of the main technological trends of the twenty-first century, neglecting what is going to happen in the fields of genetics, energy, transport and many other areas of innovation.

© The Author(s) 2019
J.-H. Lorenzi and M. Berrebi, *Progress or Freedom*,
https://doi.org/10.1007/978-3-030-19594-6_4

This confusion and this difficulty in understanding the future produce the kind of soothing, simplistic arguments, far from reality, that we are confronting today. History teaches us, however, that labour has undergone transformations before, which brought changes and reactions that would not look out of place today. Yet we settle for an over-optimistic outlook. We imagine a society awash with universal knowledge, made up exclusively of entrepreneurs, even though, paradoxically, today's society produces more unskilled people than ever. The labour market is shattered: pitting a minority of switched-on, highly skilled people against the majority of others, who settle for tasks which need no specific qualifications. But we must look to the future and imagine that this very naïve vision of a technological twenty-first century will happily become more nuanced from now on, creating highly differentiated professional opportunities and pathways for individuals. But we still need to rid ourselves of those wildly unrealistic visions of the future and reject technological illusions. We must look at the reality that is today's labour market and imagine how it could be organised, along with a completely renewed social security system, in years to come.

## 4.1   Back to the Past

Before the first industrial revolution, textile work in Great Britain was very widely carried out by the wives of poor peasants, in their homes. This was called the "putting-out system". It was a completely decentralised production system, where the work was mainly carried out in the domestic sphere, or sometimes in rural workshops, within a completely decentralised production process. The labour organiser provided the workers with raw materials as well as production tools. The organiser would collect and then market what they produced. These goods were not destined for local markets, but to supply markets in other regions or countries. Strict rules governed the products' characteristics, to provide the uniformity demanded by mass production.

This system was in use in the textile, shoemaking and locksmithing sectors in Great Britain and the United States at the beginning of the eighteenth century. But it declined rapidly after the invention of the sewing machine in 1846, particularly when poor peasants were forced to move to cities after the Enclosure Laws and provide the manpower for what would become the modern industrial factory system. The "putting-out" system was well-suited to the period before the rural exodus, because the workers did not have to leave home to carry out their production tasks. The workers enjoyed some flexibility, balancing agricultural and domestic tasks with their textile production work, the latter obviously being more intense during the winter.

What does this remind us of? Simply, of teleworking. The "putting-out system" consisted of outsourcing a part of the work carried out during a manufacturing process to sub-contractors working from home, introducing a form of what is now called teleworking or remote working. Of course, the use of loans, payment in kind and advances in goods meant the workforce was entirely subject to the work rates imposed by the owner-merchants. In fact, this *Verlagsystem,* or truck system, was practised in Flanders from the thirteenth century, by Florentine drapers in the fourteenth century and spread to England, Germany then in France in the seventeenth century, particularly in the lace-making industry. This piecework then evolved into the "sweating system", a form of intense exploitation which would be developed throughout the nineteenth century.

Of course, today's conditions are nothing like the suffering we can imagine went on in that pre-industrial world. But nevertheless, there is currently a return to some forms of in-work isolation, of separating the individual from organised labour, which strangely resemble what came before. What fundamental use are these lessons from the past to us? They simply clarify and confirm that today's payment arrangements must be strictly separated from those of the past and from the highly individualised payment system.

There is a good reason to believe that these largely fanciful visions, about knowledge and entrepreneurship developing all forms of individualism, will get stronger, without anyone really suspecting that the labour market is gradually abandoning all the territory it conquered in the twentieth century.

## 4.2    The Dream of Universal Knowledge

We are talking about an old tradition, or more accurately a utopia which reaches across time, strengthened today by a school of thought driven by Amartya Sen in what he calls the "Capabilities approach": that the development of knowledge is essential, because the final objective of an individual's development is his or her freedom. In this sense, knowledge and widespread access to it are one of the foundations of freedom. We can easily imagine this as an objective, but the leap from that to making it an accessible dream defines the whole massive illusion of the early twenty-first century. We assume that digital technology will guarantee everyone instant and unlimited access to a vast knowledge, independent of their financial means or place of study.

This fantasy of universal knowledge is not new though. We can trace its roots as far back as the Mesopotamian civilisation. The great Assyrian King Assurbanipal, who proclaimed himself king of the universe, ruled his land with an iron hand, but was also a man of letters. He had the first library worthy of that name built at Nineveh, which was designed to preserve all the knowledge—cuneiform literature—of his era. More than 20,000 clay tablets were preserved there, some about mathematics, magical medicine, or philosophy, others on prophecies, omens, predictions, religious rituals and even administrative texts. This was the library which housed the text of The Epic of Gilgamesh. For Assurbanipal, power did not only lie in arms, but perhaps even more in learning: in knowledge, which was a sign of closeness to God. Of course, this knowledge was not accessible to his subjects, but the valuing of knowledge was a first step towards the dream of universality.

Two other extraordinary figures, King Ptolemy and his son, Pharaoh Ptolemy II, built and developed the legendary library of Alexandria, which was designed to gather all of human knowledge in one place. Everything that existed in tablet or papyrus form was brought there by boat from all over the known world, to expand the original 400,000-volume collection. But it was under Ptolemy II's reign that the library took on new dimensions. Led by the great scholars of the age, it owed its fame to the immense task of the copying and translation of the most famous and precious texts, such as the Hebrew Pentateuch, which became the Septuagint in Greek. More recently, in the middle of the eighteenth century, we should recall the colossal task of compiling *L'Encyclopédie*, directed by D'Alembert and Diderot.

What should we take away from these examples from the past, in relation to what the software wizards and Internet bosses say they are giving us? Firstly, that knowledge is an eternal sign of power, but that its universality remains completely relative: The Library of Alexandra could gather the sum of the works of the human mind because their number was relatively reduced compared to today and because it left out entire civilisations. Secondly, that it was reserved for a small elite of literary people, who could read and write and had mastered the dominant language of their times. How can we compare the initiatives led by the solemnity of Aristotle's Pupil, Demetrius of Phalerum, for the Library of Alexandra, with the mass of information available on the Web?

We have no difficulty in imagining that digital technology will guarantee everyone instant and unlimited access to vast knowledge, irrespective of an individual's financial means or their place of study. The tools which support that Utopia today are many and effective: search engines, digital encyclopedias like Wikipedia and massive open online courses (MOOCs) which are seeing rapid growth. Many successful examples are restoring meaning and credibility to this fascinating goal. How satisfying it is to learn that a baker in Bangladesh, Sharmeen Shehabuddin, set up a successful small business thanks to a Wharton entrepreneurship course, or that a 16-year-old from Ulaanbaatar in Mongolia was invited to study at MIT in 2013 after he was spotted on an online course from that prestigious university. Transhumanists like Ray Kurzweil and Laurent Alexandre even imagine that from 2035, we could graft computer chips to our brains to connect to the Internet and guarantee ourselves immediate and automatic access to information.

Returning to a more classic view of access to knowledge, we can see that in order to build such an education system, six principles (Parrou 2016) must be respected. Firstly, education must of course be open to all. Secondly, it must be highly personalised: the clear ambition is to enable everyone to use programmes which are perfectly suited to them. The third principle, of communication between peers, means that the division between teacher and taught becomes rather less distinct.

The other characteristics are much more mundane: that learning must have a strongly practical basis, that is should be continuous and lifelong and finally, that it should be spread via a mixture of online and offline courses. In short, human relationships do not completely disappear, but the boundary between the two forms of education is not firmly established. The anomaly in this design for an education system for the coming years is that it is completely unrealistic. Everything in it is tailor-made and adapted to every individual, while the latest UNESCO report (UNESCO 2016) (which reminds us how essential education is to enabling sustainable development) highlights: "the remarkable gaps between where the world stands today on education and where it has promised to arrive as of 2030". This demonstrates that far from the dream of such convenient, overabundant knowledge which is so beneficial to everyone, there is a reality which can be really wretched. This report rather underlines how far we are from a real, practical application of the utopia of universal knowledge. But there are, above all, the new knowledge vectors offered by the NITCs. Far be it from us to underestimate the importance of the creation of MOOCs. That will undoubtedly bring on a triple revolution: technological, with the digitisation of knowledge; economic, by ending the uninterrupted growth in the costs of the higher education sector, with its zero productivity gains; and educational, by renewing learning through joint knowledge-building which includes the students.

But several observations should be made about MOOCs. First of all, let's examine the myth of a virtual university. Many trials have shown that they cannot entirely replace face-to-face teaching. As we saw earlier, the preferred solution is to develop a hybrid programme of studies offering a choice of courses from a number of platforms, as some universities already do. One of the limits of the move online is the very nature of knowledge, which cannot be reduced to mere information. There is another, more critical observation relating to the democratisation of access to knowledge. Opening MOOCs theoretically makes them accessible to everyone, but it does not eradicate the motivation needed to learn, in fact it even increases the need for it. We cannot be sure that our new world of knowledge will be more attractive to greater numbers than today's world is. As universities increase their audience, they attract increasingly less motivated students and must expend more effort to get them to degree level. MOOCs will therefore, according to Tyler Cowen (2013), encourage the emergence of a "hyper-meritocracy" associated with the digital distribution of knowledge. Finally, there is the hypothesis that MOOCs will enable an increase in higher education productivity. However, the main consequence, at least initially, of the spread of online-only MOOCs in order to reduce university costs would be to lower the cost of getting a degree.

These different analyses do not invite opposition to the development of MOOCs because they are responding to a new demand for learning, without, for all that, competing with face-to-face teaching. In reality, we must compare this highly optimistic vision of the opportunities for access provided by NITCs to what Edgar Morin already stated a few years ago regarding education for the future (Morin 1999). He judged that humans who are finally free must take seven lessons on board. Firstly, to truly understand what he called "the realities of human knowledge", including its inadequacies, that is to be able to master the blind spots of knowledge, its weaknesses and its difficulties. Then it was a question of how to: "encourage a way of learning that is able to grasp general, fundamental problems and insert partial, circumscribed knowledge within them". The third essential element of this new understanding of knowledge entails teaching what the human condition is, but also a fourth element that he calls "Earth identity". He wanted to underline through this that it was necessary to: "show how all human beings now face the same life and death problems and share the same fate". Knowledge was also, for him: the comprehension of uncertainties, mutual understanding between humans and the definition of an ethical system. The extent of Edgar Morin's influence on our own approach to the analysis and understanding of technology can be clearly seen through this approach to the knowledge humans need in our complex era. Could it be said that MOOCs are destined to satisfy his demands today and tomorrow? Not entirely, because, in reality, another approach is taking over. We are witnessing the shift from a vertical transmission of knowledge by "experts" to lay people (of which MOOCs are a perfect example) to a more horizontal transmission and sharing of knowledge between contributors and users, who often overlap. The current example of Wikipedia is a good illustration of how individual contributors can take part in building a collective knowledge base. Apart from this classic example, where everyone is free to contribute to the digital encyclopedia, many trials, particularly by companies, are attempting to implement this learning, which is both individual and collective. This system is horizontal to the extent that every participant learns, but they also teach by sharing their knowledge and experiences. The results of these company education programmes show that 70% of the information learned by an employee comes from his or her individual daily work, 20% from other people and just 10% from formal courses. This approach completely calls into question traditional training and human resources policies. We can see that the passing on of knowledge will have radically new forms and content in the coming decades.

But let's not delude ourselves here. This model of substitution is, in part, an illusion, for many reasons.

First of all, access to knowledge via these new means will not miraculously solve the problem of inequality of opportunity, because that continues to depend above all on pupils' environment, in particular their family situation. The democratisation of access to knowledge could in many respects appear to be a myth: the opening of MOOCs makes it accessible to everyone, in theory. However, it does not abolish, but rather increases, the motivation needed to assimilate it. We cannot fail to enjoy the examples of success quoted above, but it is obvious they only concern cases very much on the fringes of disadvantaged populations. How representative such anecdotes are is well and truly contested by sociological studies (Chuang and Ho 2016),[1] the majority of which conclude that the users of MOOCs are, at least today, mostly graduates and part of the workforce. There are no current proposals to profoundly alter that state of affairs.

The other critique widely developed by numerous academics is that of a university which becomes completely virtual. Trials have shown that virtual education is not designed to be a complete substitute for face-to-face teaching. The essentials of teaching happen via regular interactions with teachers. Also, online courses suffer from a low rate of attendance and completion. In general, less than 10% of those registered on a virtual course finish it (Rapp 2014).

Finally and perhaps most importantly, the horizontal sharing of knowledge also poses a number of challenges in authenticating and verifying information. Today, there is a long list of terms used to describe a global crisis in the truth and reliability of information: hoax, spin, alternative facts, fake news or even post-truth. They all call into question the principle of the proliferation of information sources and unlimited sharing of knowledge, if that knowledge is not reliable and if processes of verification and authentication are not put in place.

A new human appears in this vision of universal knowledge which is not only accessible but at the very heart of new social relations. A vision of a free man or woman presents itself, rather as the citizen appeared in the age of the Enlightenment. The vision is obviously significant, but a little naïve and always carries a dangerous element which we must guard against. But this figure does not stop at understanding the world, at interpreting the past and predicting the future: he or she is free to operate as they choose, conducting life and work in a completely independent way, casting "salaried" status back into the shadows of the past.

## 4.3   The Dream of Widespread Entrepreneurship

If we needed to find a definition of this new stakeholder in an innately individualistic society, founded on each individual's capacity for initiative, we can draw inspiration from Azzedine Tounés and Alain Fayolle. They have described it perfectly, using a combination of facts and creativity.

A central figure in the first phase of capitalism, the individual entrepreneur is gradually supplanted at the end of the nineteenth century by the figure of the manager, associated with the industrialisation of production and the dissociation between business ownership and decision-making power. In the twentieth century, mass production entails the development of large, complex organisations which dilute the spirit of entrepreneurship.

The end of the twentieth century marks the return of the figure of the entrepreneur, contrary to Schumpeter's intuition. He predicted its gradual disappearance, due to the concentration of economic activities. But the entrepreneur has seen a marked comeback since 1975, due to changes in demand which are rendering the standardised production of goods obsolete. Unable to benefit from their proximity to the customer, large firms outsource certain activities in order to inject a new spirit of enterprise. The growth in the number of small and medium businesses is a huge movement in developing countries, who are transitioning from a "management economy" to an "entrepreneurs' economy" (Drucker 1985). The small business has become an institution which enjoys undivided fame; social interest in the entrepreneur, the creator and driver of his or her own actions, fills the stage.

Size is no longer an aim in itself for a business and in any case is no longer considered a necessary condition of performance. The small business, on the other hand, is acquiring a double legitimacy: social, because it carries the idea of self-fulfilment and in some cases reintegration into society; and economic, because it is seen as the spearhead of innovation and new employment.

This tendency is obviously increased by the ongoing, mostly digital, technological revolution with its many opportunities for technological and digital entrepreneurs and its ability to promote services independently on online platforms. Will the twenty-first century be the one where every worker now becomes an entrepreneur? The rise of part-time and portfolio working and the return of freelancing are also factors which are eroding the model of stable, full-time, salaried employment. But above all, the development of digital platforms means we can soon envisage a model where everyone will be self-employed, working directly for the customer and liberated from all legal constraints. But does all this correspond to a real possibility, or is it the manifestation of a society which is wondering how to integrate its youth? The answer is simple. All of this appears to be an important phenomenon in public discourse, but it is marginal when it comes to the real facts: salaried employment remains the norm and freelance working is in the minority, often supplementing a salaried activity. And when we examine the development of start-ups, or new business creations, in different countries, we realise that it is much weaker and more diverse than we imagine (Graph 4.1).

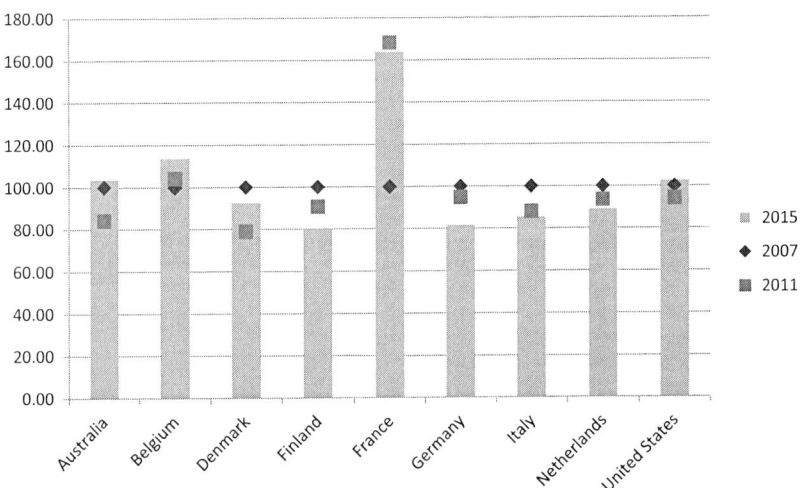

**Graph 4.1**   Development of new business creations (2007 = base 100) (*Source* OECD)

What a surprise! France, which certainly has a high level of self-entrepreneurship, holds up well. Paradoxically, we can observe a kind of stagnation of that entrepreneurial dynamism in the United States. Nevertheless, new business creations generally speeded up after the 2007/2008 crisis. Let's pursue this line of argument in the context of what we know: the return of significant growth, even though it is weaker than what we experienced before the 2007/2008 crisis. We have therefore imagined three scenarios for the years 2017–2040. Scenario 1 follows the same trend as 2007–2011, scenario 2 follows the trend of 2012–2015 and scenario 3 follows the trend of 2007–2015 (Graphs 4.2, 4.3, 4.4, 4.5, 4.6, 4.7).

We have only kept scenario 2 for France, because its legislation on self-entrepreneurs profoundly changes the forecast hypotheses. From 2002 to 2010, the number of businesses created never stopped rising, with a combined increase of 189% over the period. The increase even reached 88% from 2008 to 2010 alone, under the influence of special measures for self-entrepreneurs. Since then, the number of new business start-ups reached a growth rate of more than 500,000 units per year.

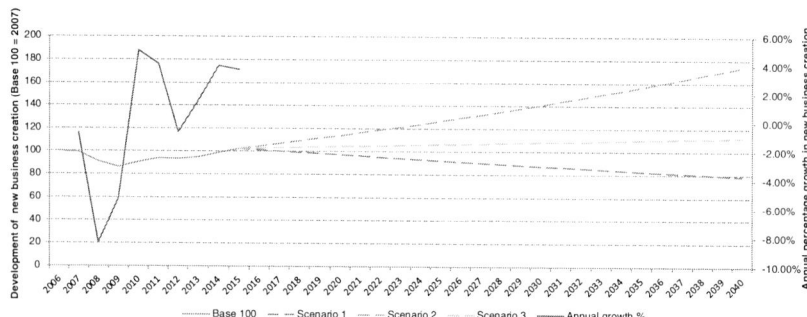

**Graph 4.2**    United States: New business creation trends from 2007 (*Source* OECD and the authors)

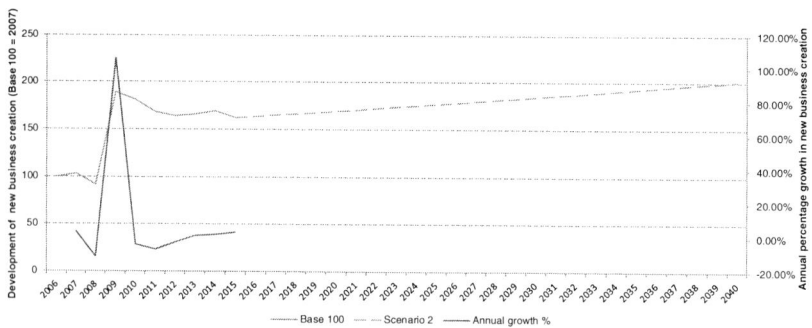

**Graph 4.3**    France: New business creation trends from 2007 (*Source* OECD and the authors)

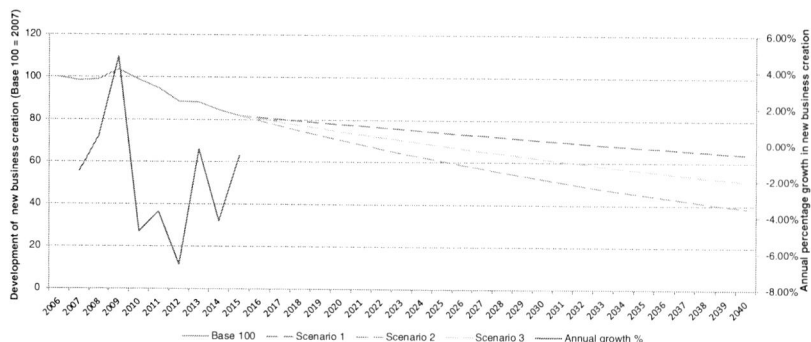

**Graph 4.4**    Germany: New business creation trends from 2007 (*Source* OECD and the authors)

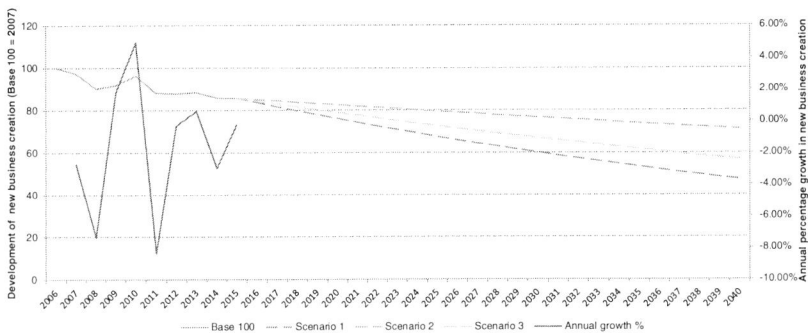

**Graph 4.5**   Italy: New business creation trends from 2007 (*Source* OECD and the authors)

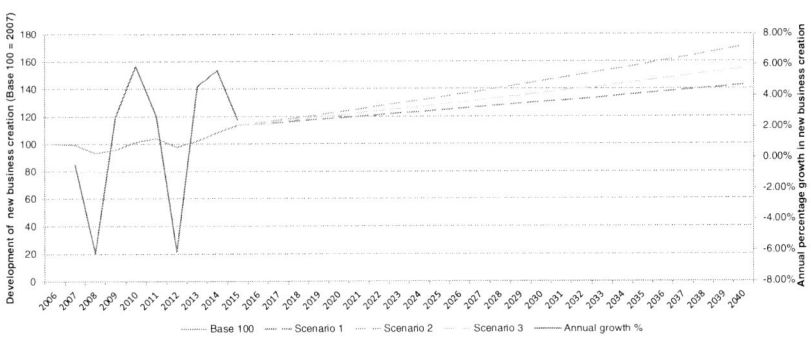

**Graph 4.6**   Belgium: New business creation trends from 2007 (*Source* OECD and the authors)

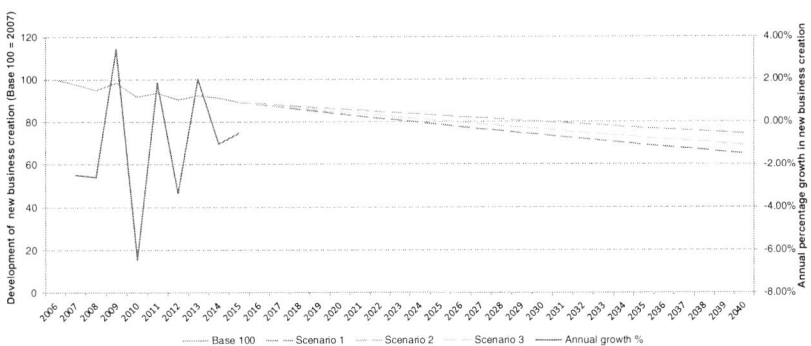

**Graph 4.7**   The Netherlands: New business creation trends from 2007 (*Source* OECD and the authors)

In France and the United States, we can see how strong the new business creation movement is, but we can also see it is not entirely key to the projected development of our economies.

What are these businesses like? Are they different today from the way they were before? Yes: they do have a more pronounced air of personal adventure, which gives them a feeling of real disruption. In France, only 4.6% of new businesses—and 8% if we include traditional businesses—have salaried employees when they launch their enterprises. That is down from 13% in 2008 to 6% in 2009. Between 2000 and 2015, the number of start-up businesses with no salaried employees has tripled while the number employing them has fallen to 38% (APCE 2015) (Table 4.1).

This is undoubtedly the sign of a real, slow but strong transformation of the labour market. This development is, in fact, a combination of different factors. Firstly, changes due to unemployment have become more frequent. Secondly, the beginning of working life is characterised by instability. In France, for example, under-30s are twice as likely to be unemployed next year as over-30s and young people are very widely hired on fixed-term contracts (France Stratégie 2016).

**Table 4.1** Business creation in France in 2015 and compared to 2014 in terms of salaried employees at launch

| Size of new business at creation | Number of creations in 2015 | Distribution of creations in 2015 (%) | Change between 2014 and 2015 (%) |
|---|---|---|---|
| A. Businesses with no salaried employees | 500,695 | 95.4 | −4.5 |
| B. Businesses with salaried employees | 24,396 | 4.6 | −7.5 |
| Proportion of B. with: 1 or 2 salaried employees | 18,842 | 3.6 | −5.1 |
| 3 to 5 salaried employees | 3165 | 0.6 | −13.5 |
| 6 to 9 salaried employees | 897 | 0.2 | −9.4 |
| 10 or more salaried employees | 1492 | 0.3 | −20.1 |
| Total number of new businesses | 525,091 | 100 | −4.7 |

*Source* APCE

A growing number of recruitments are therefore for a limited time and also for shorter and shorter periods of employment, even if the proportion of those contracts among the total number of employed people remains stable.

All of this signifies the transformation of the life cycle of work, and the dream of entrepreneurship plays a role in that.

Let's go back to our projections, (see Table 4.2) whose sole merit is that they illustrate what non-salaried employment will earn in 2040. The term "non-salaried employment" includes employers, self-employed people, members of production co-operatives and family members doing unpaid work. We have come up with projections up to 2040 following three different scenarios. Scenario 1 assumes the rate of change in non-salaried employment between 2015 and 2040 is identical to that observed between 1985 and 1995. Scenario 2 assumes this rate is identical to that observed between 1995 and 2005. And scenario 3 assumes a change in this rate identical to that observed between 2005 and 2015.

Here too, it is difficult to conclude that the disappearance of salaried work is taking over and represents a decisive trend. Of course, we must not ignore all of this, but the definitive transformation in the labour market is not happening yet. We can undoubtedly support the idea that the development of different forms of employment will be largely determined by the powerful interaction between changes in technology and

Table 4.2   Rates of non-salaried employment in %

|  | France | Germany | Italy | Japan | Spain | UK | United States | OECD |
|---|---|---|---|---|---|---|---|---|
| 1985 | 15.0 | 11.8 | 29.7 | 25.4 | 30.2 | 13.1 | 9.1 | – |
| 1995 | 10.8 | 10.7 | 29.3 | 18.3 | 25.2 | 14.5 | 8.5 | 19.2 |
| 2005 | 9.0 | 12.4 | 27.0 | 14.7 | 18.1 | 12.9 | 7.5 | 16.7 |
| 2015 | 9.7 | 10.8 | 24.7 | 11.1 | 17.4 | 15.4 | 6.5 | 15.8 |
| 2040 with Scenario 1 | 2.9 | 8.2 | 23.9 | 3.3 | 10.2 | 19.5 | 5.4 | – |
| 2040 with Scenario 2 | 5.6 | 15.2 | 19.8 | 5.6 | 5.1 | 11.3 | 4.5 | 10.8 |
| 2040 with Scenario 3 | 11.5 | 7.2 | 19.3 | 4.3 | 15.7 | 22.9 | 4.3 | 13.6 |

*Source* OECD and the authors
See Graphs 4.13–4.20

changes in behaviour, not forgetting the role of labour market institutions and of changes in social security.

Let's just remind ourselves that projections of employment by occupation do not show the kind of disruption which would cause a significant distortion in the structure of employment. Employment structure will remain affected by the same transformations that are already happening: expansion of the service sector, growth of service jobs and those linked to ageing. In France, if its economic and regulatory environment remains constant, atypical forms of employment without a permanent contract could represent at most a quarter of employment in 2027, compared to 22% today (France Stratégie 2016).

More radical disruption is not out of the question. This could be caused by the combined effects of the emergence of new models—platforms—for co-ordinating labour: a lowering of employers' salary costs and a change in regulation and practices for the most independent workers and for those who aim to top up insufficient salaries. But platforms are a perfect illustration of the difficulty of predicting the future. On one hand, the potential of the platforms model should not be underestimated, because it combines a structure which directly links service providers and customers and a guarantee of quality from users' reviews. But on the other hand, the status of workers from platforms is variable in practice because this model can just as easily be applied to those offering high value-added services as to those offering services from poorly qualified workers instead of salaried employees. The development of these forms of employment is therefore likely to make careers more irregular and incomes more volatile. Disintermediation and the precarious and portfolio nature of careers may create a need for new forms of professional relationships, particularly the reforming of workers' collectives, which can offer a framework for negotiating employment conditions and income. This takes us a long way from the liberating image of entrepreneurship.

Let's leave the conclusion to Augustin Landier and David Thesmar (2015) to put this vision of a world without salaried employees into perspective: "For several years, it has been stated repeatedly that new technology will make the salaried worker obsolete. The worker of the future will be a freelancer, collecting payments completely independently, according to the needs of his or her patrons. New technologies favour this kind of work structure because they put labour supply and demand in direct contact". Nevertheless, on the employment front, the end of the salaried worker is slow to appear. As we have seen, in the United States, the proportion of non-salaried workers has gone from 7.5 to 6.5% of total employment between 2005 and 2015, whereas in France, the proportion of non-salaried workers has risen slightly, from 9 to 9.5% over the same period.

This contradiction between the dynamism of new business start-ups and the permanence of the salaried workforce, caused by the rise in multiple working, is revealing. The new paradigm is entrepreneurship in tandem with salaried work. So, what is becoming widespread is not freelancing, but a secondary income from freelance work complementing a stable, salaried income. This change enables salaried workers to improve their skills and create new possibilities for development, while the stability of the salaried income enables them to take more risks in their freelance work.

Obviously, this vision relies on all individuals having strong skills and on the ability to fall back on unquestionable skills and knowledge. That might be what will happen in the future, but, today, it produces the opposite effect: that of proven skills on one side and routine tasks on the other. The former relies on the power of technology; the latter is seeing its qualifications whittled away by digital technology.

## 4.4   The Boom in Low Qualifications

Industrial revolutions have always followed a similar scenario: the same logic of short-term imbalance and hope of prosperity further down the line. In the initial phase, the machine takes the place of labour: not only in terms of the number of jobs, but also the qualifications that workers apply in practice. This inexorable logic leads to today's famous polarisation of the labour market, seen in all developed countries, pushing them all to two extremes: ultra-sophisticated roles, or very low-skilled tasks.

Remember, qualifications determine individuals' ability to carry out a job and to accomplish given tasks. Valued on the basis of degrees obtained and skills acquired through professional experience, they are a measure of employability for salaried workers and of effectiveness for employers looking for a skilled workforce. For about 10 years, the growth of the average grade achieved at degree level has hidden a phenomenon of polarisation of qualifications to the two extremes of the scale. Throughout the developed world, there is a boom in qualified employment requiring creativity and analysis skills and also in low-qualified jobs with no prospect of enrichment through work. This polarisation is accentuated by the decline in middling jobs whose routine tasks can easily be automated, whether they are manual or require intellectual skills.

The disappearance of these middling jobs carries another risk: that of permanently locking a growing proportion of the population into low qualifications, thus putting at risk any prospect of social mobility. In summary, the automation of many tasks raises, at least in the short term, the spectre of "technological unemployment" when machines are substituted for workers. But there is no certainty in these areas: the impact of technical progress on work remains uncertain in the long term. Contrary to the unwisely disseminated theory of the "end of work", the complexity of the relationships between employment and technology and the possibility of creating new jobs using digital technology suggest that digital technology is more likely to replace jobs than eliminate them.

The graph below gives us unequivocal results. It shows how far this bipolarisation has developed in France, the United States and many other OECD countries throughout the beginning of the twenty-first century (Graph 4.8).

This is all confirmed by a recent study carried out by the Saint Louis Federal Reserve Bank (Dvorkin 2016) which incontrovertibly demonstrates that jobs are becoming more and more concentrated in high value-added roles or in the opposite type of role, with low qualifications. "Employment in routine occupations […] has been mostly stagnant" the report says. It describes the phenomenon of "job polarization". The study differentiates so-called routine jobs from "non-routine" and roles with "cognitive" skills from "manual" ones. Only non-routine, cognitive jobs have seen a real increase in the last thirty years, while the other categories have stagnated, or indeed decreased. The report says this tendency will only continue in the coming years.

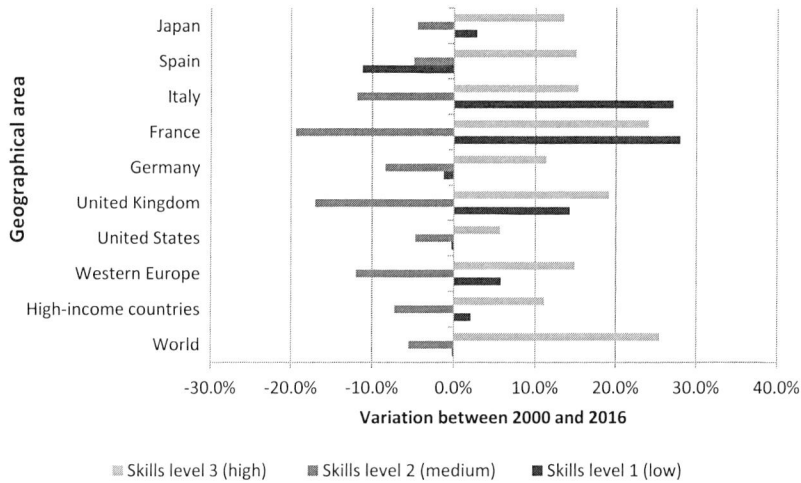

**Graph 4.8** Variation in the distribution of employment based on skills level between 2000 and 2016 (*Source* International Labour Organisation and the authors)

The issue then is to understand whether dynamic sectors like health, education and maintenance will compensate for their job losses, most of which are for people with middle-level qualifications. If most economists anticipate an elimination effect—in other words, a broadly negative balance between the destruction and creation of jobs—others point to a substitution effect. Thus, Harry Holzer (2015) talks about "two middles" to describe those intermediate jobs: he draws a distinction between jobs which are in decline, such as construction, and occupations in dynamic sectors, such as those related to health. In his opinion, the latter sector is suffering from a shortage of available workers and therefore must be at the heart of future employment strategies in the years ahead. But while that development is most certainly desirable, it is marginal compared to the problems to be faced.

In order to explore the future of this trend and its evolution towards bipolarisation, we must understand its origins. For Carl Frey and Michael Osborne, the automation which is at the origin of this movement can be measured according to two criteria: the repetitive/routine nature of the cognitive or manual tasks carried out and how far technology must progress in order to effectively automate that task. They conclude that 47% of jobs have a high probability of being automated, 19% a medium probability and 33% a low probability (Frey and Osborne 2013).[2] This convinces us that bipolarisation is increasing.

For David Autor, technological advancement is just one the factors causing bipolarisation. It can be explained by many others, such as: international trade and the relocation of some jobs, factors linked to labour market institutions and to individual countries' labour relations,[3] particularly via the role of unions and minimum wage legislation. Other factors include sociodemographic changes and the growth of the service sector of the economy. This makes for a much more complicated picture, with a more uncertain outcome.

What will the future be? No one has firm convictions. We simply know that automation is making intermediate manual and intellectual jobs disappear, if they are characterised by routine tasks which are easily automated. In order to illustrate this statement, we have drawn up simulations which extrapolate the bipolarisation phenomenon observed between 2000 and 2016. This shows in a purely illustrative manner how following this trend profoundly transforms the labour market (Graphs 4.9, 4.10, 4.11, 4.12).

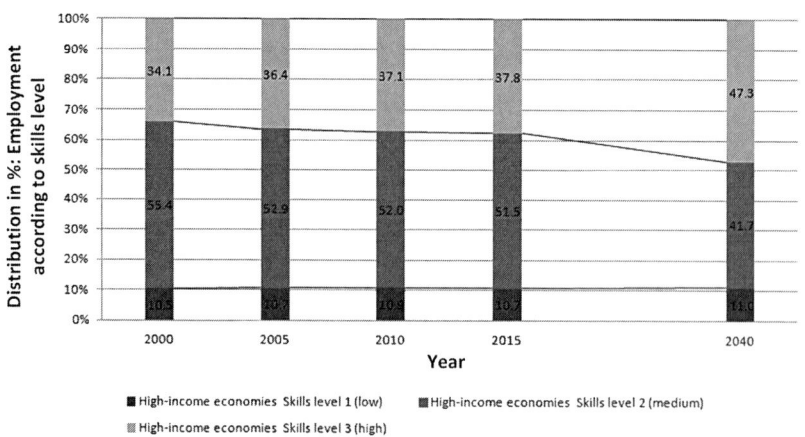

**Graph 4.9** World – High-income economies: Employment according to skills level (*Source* International Labour Organisation and the authors. See also graphs 4.20–4.23 in the appendix)

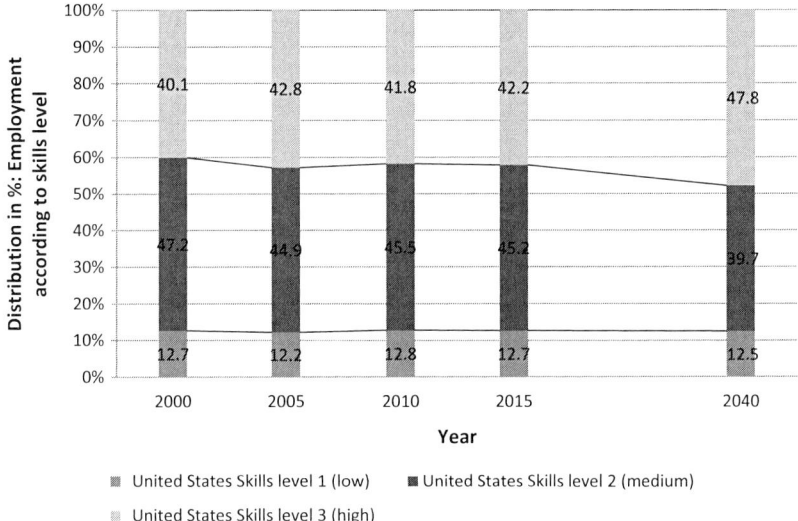

**Graph 4.10**  United States: Employment according to skills level (*Source* International Labour Organisation and the authors)

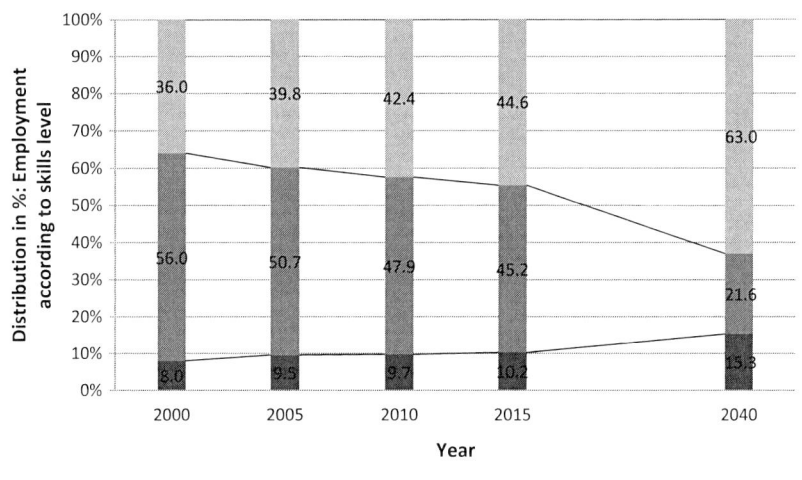

**Graph 4.11**   France: Employment according to skills level (*Source* International Labour Organisation and the authors)

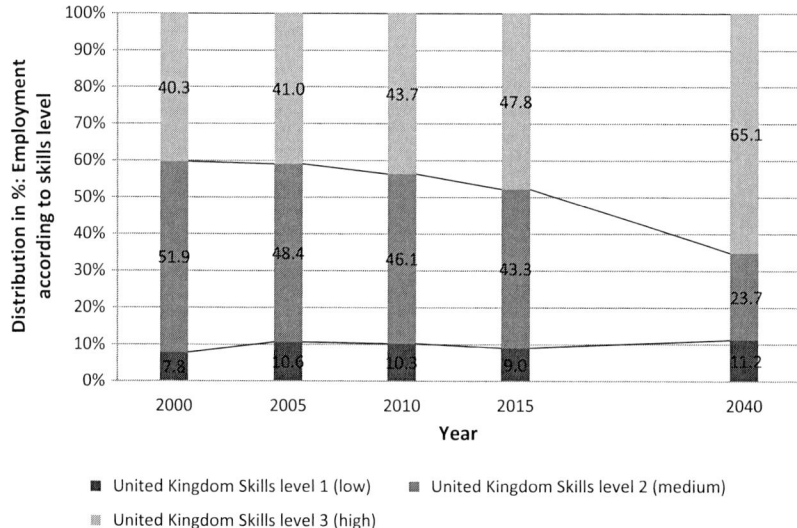

**Graph 4.12** United Kingdom: Employment according to skills level (*Source* International Labour Organisation and the authors)

The graphs are mere representations all of the hypotheses which are up for discussion. Nevertheless, they reveal potentially very significant shifts which are turning salaried employment on its head, while we know that on the ground, salaried status looks as though it will remain the principle status of employment.

In conclusion, while the polarisation of qualifications and the disappearance of many routine jobs are recognised in these early years of the century, uncertainty reigns over the coming impact of technology on employment and on the creation of new jobs in the medium term. In short, there is total uncertainty. It simply highlights the absolute necessity of rethinking the challenge of protecting individuals whose professional careers are now fragmented, within a labour market which, as we have seen, has been blown wide open.

From this perspective, it is wage relations which must be adapted, with the explicit aim of protecting the individual from life's accidents.

## 4.5   New Wage Relations

If the term "social forecasting" has a meaning, it is very much in this area that we can apply it. It is rare to find instruments for social dialogue and protecting the individual at work which are truly adapted to the complexity of a world of work in mid-transformation. And what is true today will obviously be much more so tomorrow.

Social relations in France are still experienced in the context of a social contract drawn up at the end of the Second World War by the National Council of the Resistance. This contract was entirely designed for a Fordist organisation of labour under the social and economic conditions of "*les Trente Glorieuses*".[4] Until the early 1970s, the welfare state was the cornerstone of this compromise partnered with industrial capitalism: a balance between the interests of the market and those of the world of work. It was, in fact, the culmination of reforms inspired by social democracy, designed to regulate the market using rights and protections which benefitted its central figure, the salaried worker. It is in fact salaried work which brings together and creates relationships, and it is the salaried worker who is protected by collective insurance schemes against work accidents, illness and unemployment.

The 1945 social pact is, however, designed for and in a full-employment society driven by growth—the average annual rate was 5.3% from 1945 to 1974—and its faults show when the oil crisis hits in 1970 and when mass unemployment appears. The assaults on this model do not remain purely financial. The social pact actually finds itself swept up in the turbulence of a huge reorganisation of society which goes well beyond the socio-economic sphere. After having analysed the social security crisis, Pierre Rosanvallon (1981) very quickly heralded the advent of a new social debate and the intellectual and moral reconstruction of the initial pact.

In fact, the list of shocks for which the welfare state is not prepared, due to its lack of compatibility with globalisation, is a long one: the rise of individualism, the unprecedented decline in representation by collectives such as unions, the economic crisis (which feels like the twilight of an industrial society whose traditional methods of regulation are exhausted) and the crumbling of the salary-earning society, replacing an integrated educational and professional system with insecure careers, are developments which also rock the post-war contract down to its foundations. As Daniel Cohen (2006) reminds us, the post-industrial society emerged from disruptions which have turned our world on its head over the past thirty years: technological, with the Internet; social, with the new organisation of work; cultural, with the promotion of the individual; and financial, with the famous and dangerous "genie out of the bottle" and the new globalisation. And we must of course add to this the highly significant observation that longevity will continue to increase, at least for the present.

This rapid overview of the 1970s and 1980s shows how sudden the change was and how much the world which emerged is marked by uncertainty. If the "no gods, no masters" baby-boomer generation of 1968 was hardly touched by this disturbance, the same is not true for the generations which followed, who live, like survivors of the shipwreck of the industrial society, in a state of "social insecurity" as Robert Castel calls it. From the beginning of the 1990s, the sociologist François Dubet highlighted this young generation in difficulty, bearing the cost of something that is by no means a readjustment, but in fact, the disqualification of a social structure, corresponding to a changed economic landscape, whether that applies to integration, social relations or means of collective action. This generation is already accumulating periods of enforced idleness, of small "gigs", a kind of continuous pressure which, in the end, turns them into an age group haunted by the perception of a world dominated by uncertainty. At the other end of the scale, there are seniors lacking adequate training who are joining the ranks of the long-term unemployed. The 1945 pact is excluding people, rather than integrating them: this is an unsustainable anomaly.

What must be well understood is that wage relations fundamentally rely on a model of social security decided at a collective level, with rules which apply to the company under a labour contract. In this setting, this contract fixes specific rules for remuneration and protection of the individual against life's accidents. These can relate to health, life and critical illness cover, training, or possibilities for career development within the company, which is the most difficult one to achieve. It is this very wide definition which enables the relationship between employer and employee to be specified at the collective level and at the company level. Now everyone has their own vision of the future. Some people think that the dialogue between these two groups will transfer to the individual themselves, who will be his or her own employer from now on. Others (who are more realistic, as we were in the preceding paragraph) think the opposite: that the rules for an individual entrepreneur will approach those for a salaried employee, until they incorporate them. We believe that the latter scenario is more likely in the coming decades, but that is not the key point. The issue is to really understand what the power relationship will be and how it will express itself between employers and employees. Our hunch is that the protections we have now, which are so reassuring for those who enjoy them, will be gradually called into question, thus putting an end to this period of Fordism, mass consumption and the middle classes which is so appreciated by all.

Are we then witnessing the beginning of the end of our social security system, which is still so precious to us? We certainly are. Since it was based on a strict separation between salaried status and that of the freelancer, it no longer offers an adequate response to these new, developing, forms of work which straddle the two statuses. Why are we so pessimistic? Because the main risk is that a growing proportion of workers no longer benefit from the advantages of a salaried worker—social security, benefits and rewards—but at the same time do not have the flexibility which is characteristic of self-employed status. How then can we make individual, professional career paths more secure? Without a doubt, we must invent ways of adapting to these changes, by creating an intermediary status which guarantees everyone most of the protections enjoyed by salaried employees. Lifelong training is surely the key lever for a fundamentally renewed social security, quite simply because the impact of technological progress on the labour market, at least for the next few years, will be, as we have seen, an unparalleled separation between those who are qualified and those who are not. And the aim of a caring society must be to enable the unqualified to escape from a condition of servitude. All of this is already happening right now.

Seth Harris and Alan Krueger (2015) consider that with new technology, particularly in the United States, a new category of worker has appeared, halfway between the status of employee and freelancer: the "independent contractor". This category can choose, as freelancers can, its working hours or days and choose which intermediaries it works for. Meanwhile, those intermediaries—a classic example is Uber—have, on the other hand, a not insignificant control over their workers, particularly by fixing the prices charged or by being able to exclude a person from the service. These characteristics are those of salaried employment. Ultimately, this category does not benefit from the major advantages of salaried employment—social security, benefits, and rewards—or from freelancers' room for manoeuvre. In order to guarantee access to most of the protections enjoyed by employees, such as the right to unionise, or a guaranteed regular income, Harris and Krueger recommend the creation of a specific status for these hybrid employees known as "independent workers". This status would guarantee access to most of the protections which employees enjoy, such as the right to unionise and a regular income.

There is no doubt that currently there is a risk of marginalisation of a section of the working population because their social security is not fit for purpose. Some reforms have already been brought in by some countries. In Italy, for example, with the creation of "project collaboration contracts" and above all in Spain, where the status of "independent worker" was established in 2007. The latter includes, specifically, a foundation of common rights, a balance between family and professional life, protection from risks, and collective rights, which are: membership of a union or managerial organisation and in particular the freedom to create, or belong to, a professional organisation.

And obviously there is the UK, where two intermediate categories have been established between employees and freelancers. There are: "workers", people who work for an employer without being under its authority, who benefit from a minimum wage, working hours and leave. There are also "self-employed workers". They own their equipment, are responsible for organising and carrying out the work and also assume a portion of the financial risk. They enjoy health and safety rights.

But the topic does not end there. If we imagine extending this social security system to this new category of work, we will have to find suitable financing methods for this social contract. And the future of this issue is very hard to grasp. There are in fact three different issues. First of all, we have to determine the financing base, but above all the macro-economic consequences that it could have on growth and competitive-ness. Then, as Jérôme Glachant (2014) reminds us, we must "identify the winners and the losers (in terms of age and qualifications) from changes to the funding base". Finally and most importantly, In order to understand what the new wage relationship will be when it is extended to new categories of workers, we must "consider the link between universal, obligatory systems and more private, individual systems". But we must also consider new elements of this newly thought-out social security system and of the salaried relationship which we can infer from it. Alain Villemeur (2012) demonstrates very well how much social security, which enables "individuals and households to face the consequences of social risks, that is to situations which could cause a lowering of resources or a rise in spending (old age, illness, disability, unemployment, family responsibilities, etc.)" is going to change very profoundly and spread to as-yet-unchartered territories. And beyond that, social relations will obviously be changed deeply, with an effective divide between the compulsory regime and the private system, which is totally new. Alain Villemeur, for example, focussed on education for small children:

> Let's examine this area of social security which was first investigated in the 1990s. Spending on education and reception of young children (EAJE)[5] of less than three years of age (usually aged over 1 year) plays a very particular role within education spending. Schooling from 3 to 6 years of age being almost universal in developed countries, this question arises for the under 3s.

"Enrolling a child in an EAJE centre is not just beneficial to the child, but also to the family: mothers can work and contribute to the family income". There you have one of the many areas to explore, in the hope that the beneficial aspects of these innovations will outweigh the financial costs incurred. "How to explain such positive repercussions? For the Economics Nobel laureate James Heckman, early childhood constitutes a unique moment for investing in human capital, the attention and care paid to small children being of enormous importance, taking account of the exceptional neurological development they undergo at that stage".[6]

We can clearly see that from this very positive perspective, the wage relationship is transformed, because collective investment has a direct influence on salaried employees, their productivity, companies' results and by their very nature, the relationships between employers and employees.

Not just in France, but throughout Western countries, the labour market is undergoing unprecedented shocks. The enormous difficulty we are confronted with is to grasp the importance of the changes underway, to imagine a totally different configuration for the organisation of labour, to come up with measures designed for the next twenty years, all the while protecting the individual and trying to restore fulfilling content to most human activities. That will be the basis of the new wage relations.

## Notes

1. For example the joint study by Harvard and MIT: Chuang I & Ho AD (2016) *HarvardX and MITx, Four Years of Open Online Courses.* Available via https://papers.ssrn.com/sol3/papers.cfm?abstract_id=2889436.
2. Benedikt Frey C & Osborne MA (September 2013) The Future of Employment: How Susceptible Are Jobs to Computerisation? Oxford Martin Programme on Technology and Employment. https://www.oxford-martin.ox.ac.uk/downloads/academic/The_Future_of_Employment.pdf.
3. Minimum wage, flexible employment rights, etc.
4. The 30 years of strong economic growth after the Second World War in France.
5. L'éducation et l'accueil des jeunes enfants (EAJE): The French equivalent of UK Early Childhood Education and Care (ECEC).
6. Ibid.

## Appendix

See Graphs 4.13, 4.14, 4.15, 4.16, 4.17, 4.18, 4.19, 4.20, 4.21, 4.22, 4.23, 4.24.

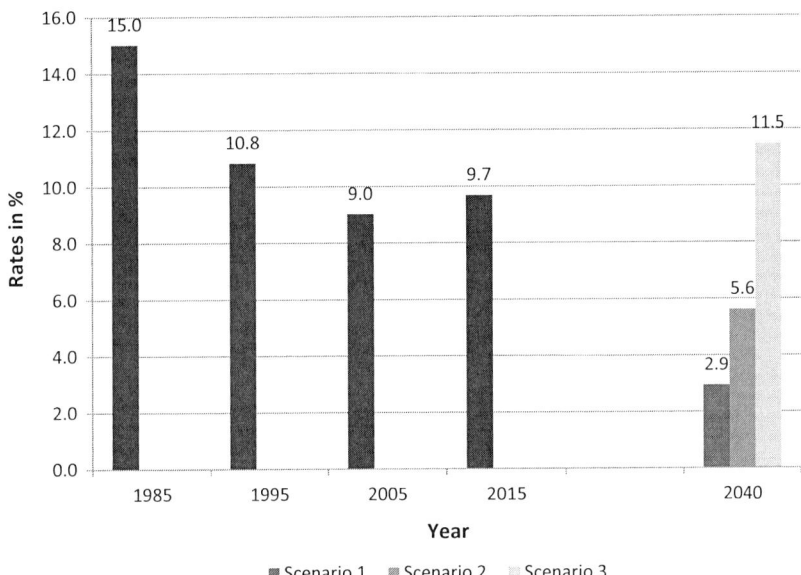

**Graph 4.13** France: Rates of development of non-salaried employment as a % of total employment (*Source* OECD and the authors)

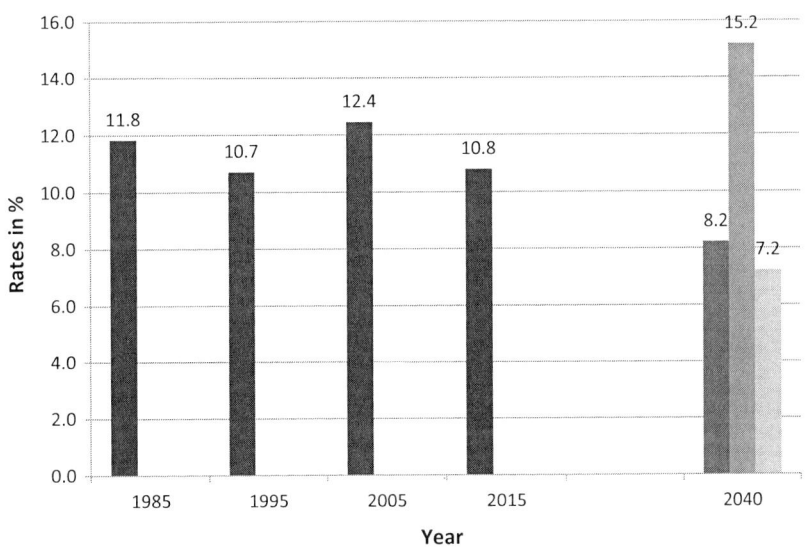

**Graph 4.14** Germany: Rates of development of non-salaried employment as a % of total employment (*Source* OECD and the authors)

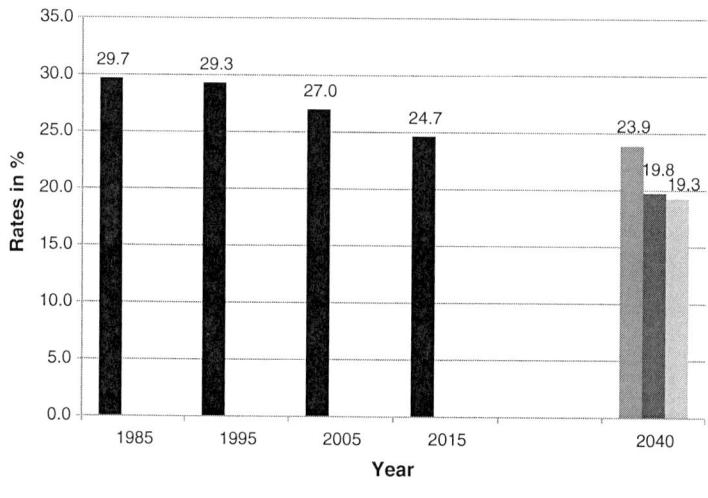

**Graph 4.15**   Italy: Rates of development of non-salaried employment as a % of total employment (*Source* OECD and the authors)

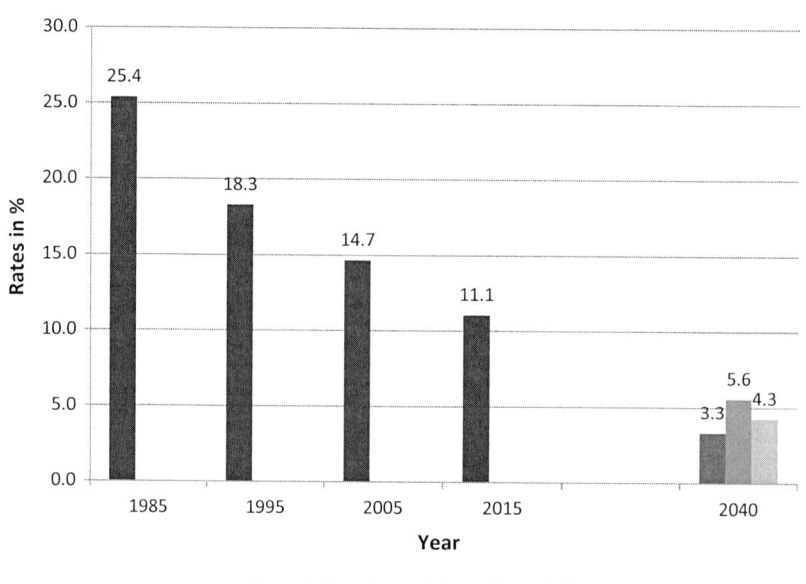

**Graph 4.16**   Japan: Rates of development of non-salaried employment as a % of total employment (*Source* OECD and the authors)

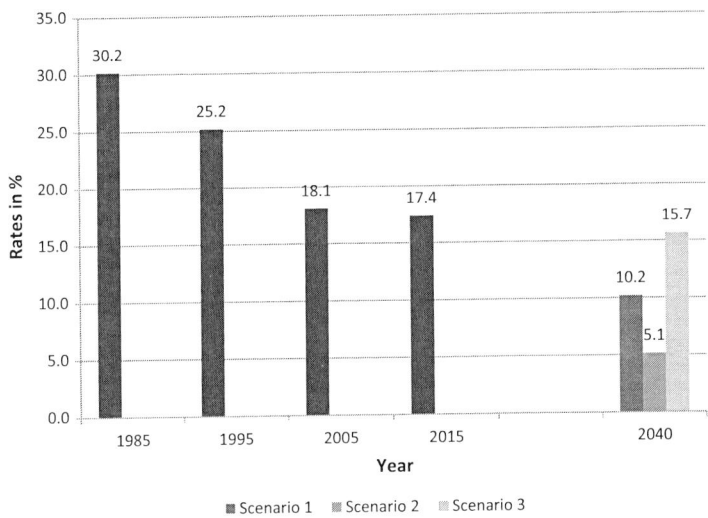

**Graph 4.17**   Spain: Rates of development of non-salaried employment as a % of total employment (*Source* OECD and the authors)

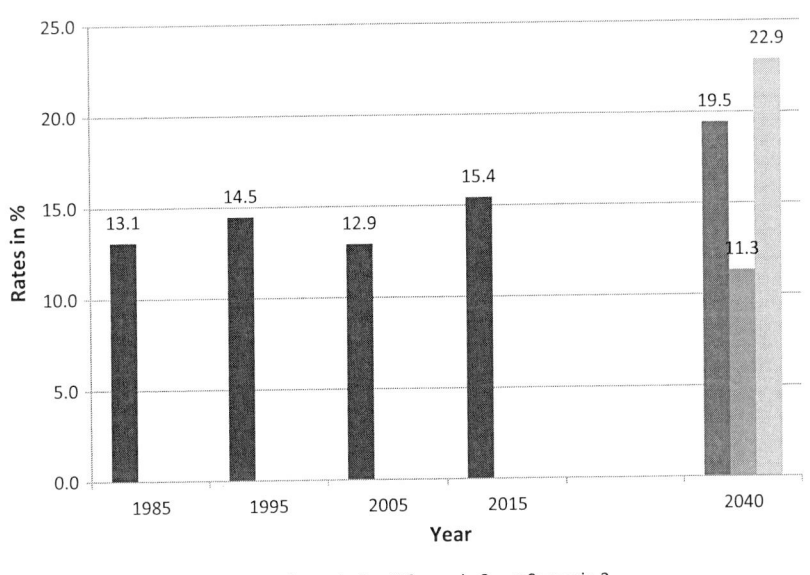

**Graph 4.18**   United Kingdom: Rates of development of non-salaried employment as a % of total employment (*Source* OECD and the authors)

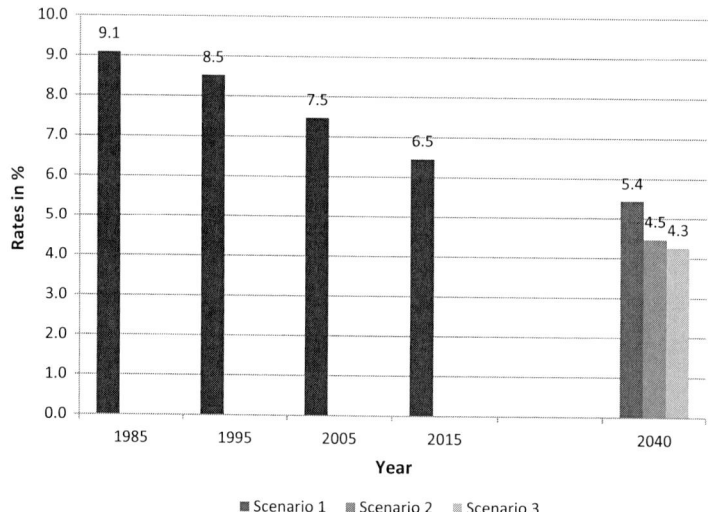

**Graph 4.19** United States: Rates of development of non-salaried employment as a % of total employment (*Source* OECD and the authors)

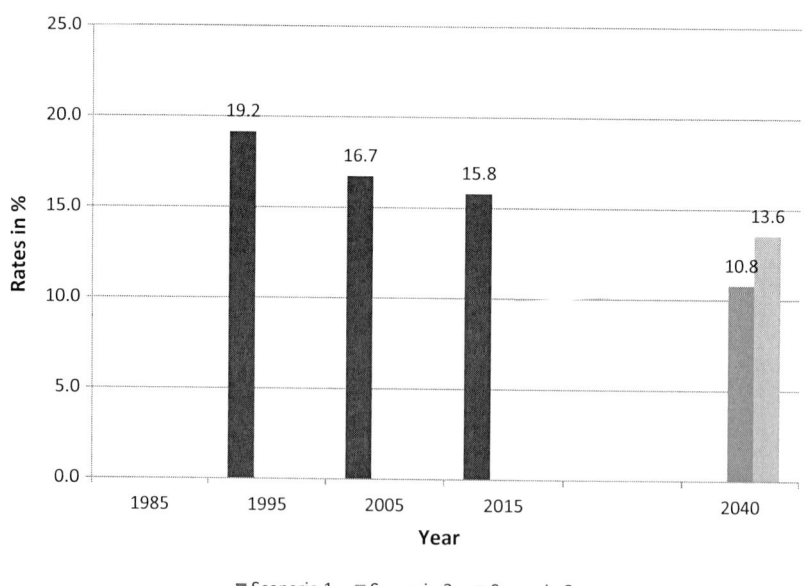

**Graph 4.20** OECD: Rates of development of non-salaried employment as a % of total employment (*Source* OECD and the authors)

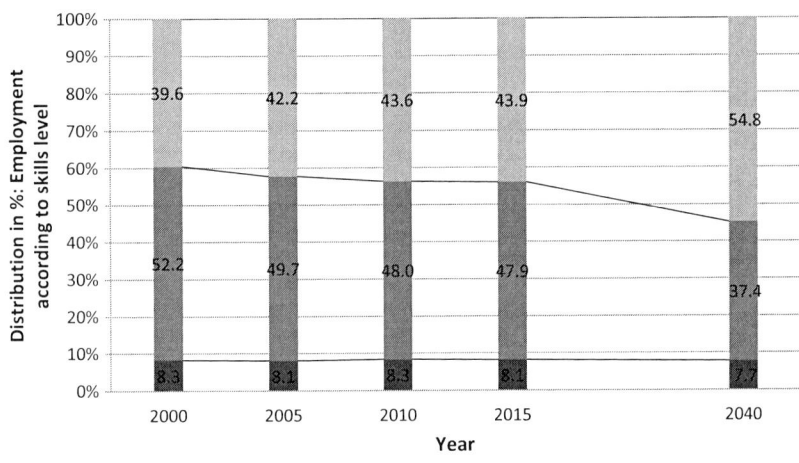

**Graph 4.21** Germany: Employment according to skills level (*Source* International Labour Organisation and the authors)

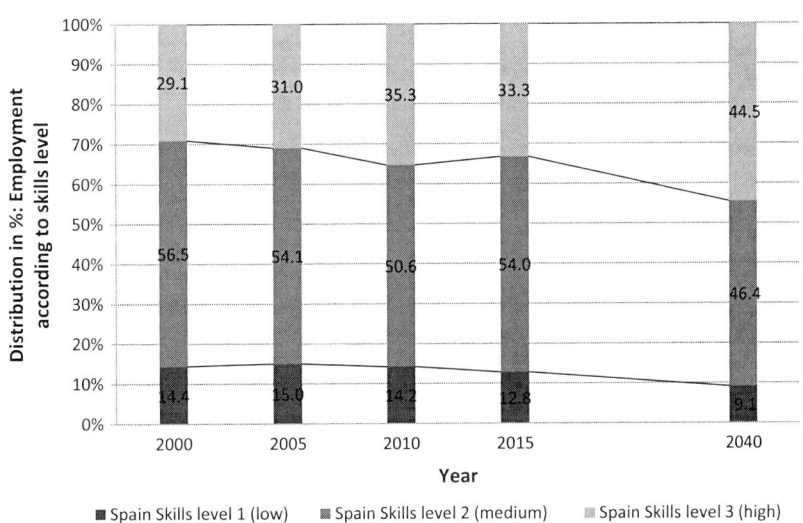

**Graph 4.22** Spain: Employment according to skills level (*Source* International Labour Organisation and the authors)

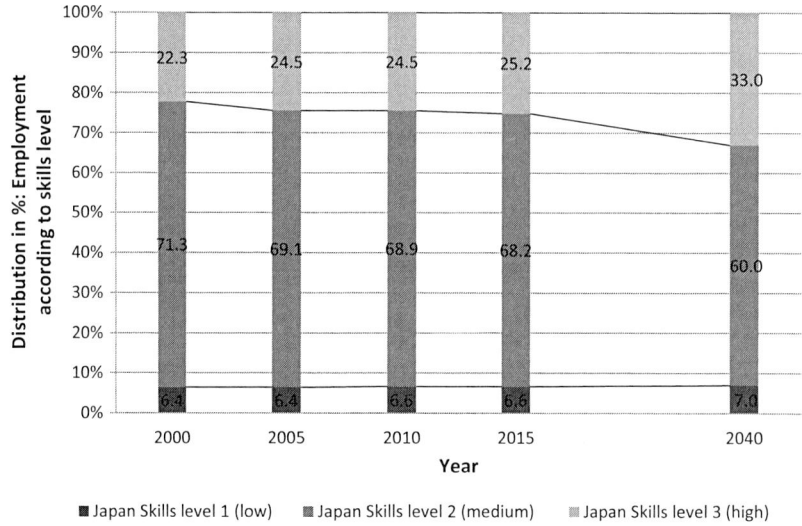

**Graph 4.23** Japan: Employment according to skills level (*Source* International Labour Organisation and the authors)

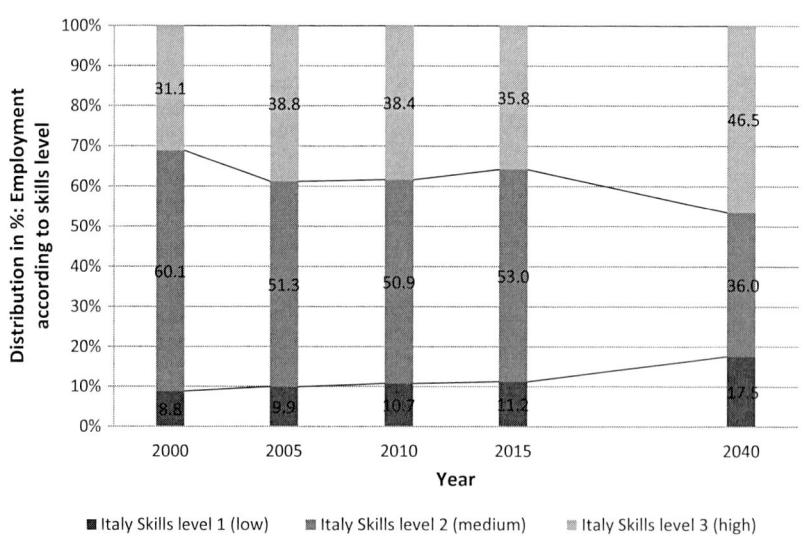

**Graph 4.24** Italy: Employment according to skills level (*Source* International Labour Organisation and the authors)

# Referenes

APCE. (2015). *Conjoncture annuelle—La creation d'entreprises en France en 2015* [Agency for Entrepreneurs, Annual Survey—Business Creation in France in 2015]. Available via https://www.afecreation.fr/pid251/observatoire-de-la-creation.html.

Cohen, D. (2006). *Three Lectures on Post-Industrial Society* (W. McCuaig, Trans.). Cambridge and London: MIT Press.

Cowen, T. (2013). *Average Is Over: Powering America Beyond the Age of the Great Stagnation*. New York: Dutton.

Drucker, P. (1985). *Les Entrepreneurs*. Paris: Édition Lattès.

Dvorkin, M. (2016). *Jobs Involving Routine Tasks Aren't Growing*. Federal Reserve Bank of Saint Louis. Available via https://www.stlouisfed.org/on-the-economy/2016/january/jobs-involving-routine-tasks-arent-growing.

France Stratégie. (2016). *Nouvelle formes de travail et de la protection des actifs* [New Forms of Labour and of Workforce Protection]. Available via https://www.strategie.gouv.fr/publications/20172027-nouvelles-formes-travail-de-protection-acti

Glachant, J. (2014). *La protection sociale: comment la financer?* [Social Securi How Do We Finance It?] Risques, no. 98.

Harris, S. & Krueger, A.B. (2015). *A Proposal for Modernizing Labour I for Twenty-First Century Work: The "Independent Worker"*. The Hami Project, Washington, DC. Available via http://www.hamiltonproject. assets/files/modernizing_labor_laws_for_twenty_first_century_work_l ger_harris.pdf.

Holzer, H. (2015). Job Market Polarization and US Worker Skills: A T Two Middles. In *Economic Studies at Brookings*. Available via https:// brookings.edu/wp-content/uploads/2016/06/polarization_jobs_polic zer.pdf.

Landier, A., & Thesmar, D. (2015). *Non, la fin du salariat n'est p demain* [No, the End of the Salaried Worker Is Not for Tomorrow Les Echos. Available via https://www.lesechos.fr/07/10/2015/l fr/021386586998_non--la-fin-du-salariat-n-est-pas-pour-demain.h

Morin, E. (1999). *Seven Complex Lessons in Education for th* UNESCO Publishing. Available via https://unesdoc.unesco.org/ar pf0000117740.

Parrou, G. (2016). *6 caractéristiques de l'éducation du future*. blog rooms.com. English version available via http://blog.openclassr en/2016/10/28/6-characteristics-of-future-education/.

Rapp, L. (2014). Les MOOCS, révolution ou disillusion? In *Les Notes de l'Institut, Institut de l'Etnreprise*. Paris. Available via https://www.institut-entreprise.fr/archives/les-moocs-revolution-ou-desillusion.

Rosanvallon, P. (1981). *La Crise de l'État-providence* [The Crisis of the Welfare State]. Paris: Seuil.

UNESCO. (2016). *Education for People and Planet*. Available via https://en.unesco.org/gem-report/report/2016/education-people-and-planet-creating-sustainable-futures-all.

Villemeur, A. (2012). *La Protection Sociale: un investissement pour notre avenir* [Social Security: An Investment for Our Future]. Paris: Seuil.

# 5

# Human Genius at the Controls

Every era in human history has the same dream, the same fantasy and the same madness: wanting to put an end to anything that cannot be understood or expressed. In short, to make human knowledge triumph! What was the Garden of Eden, if not the first appearance of the man-god who knows and masters the past, present and future? But in these early years of the twenty-first century, we are perhaps closer than ever, if not to realising that dream, at least to approaching absolute knowledge. It is true that more than ever before, we feel we are unravelling the mystery of human creation. We hope to master genetic engineering; we feel as though nothing can escape us, thanks to large-scale data processing.

It is this crazy ambition that we will discuss here, or more exactly, some examples of the uncontrolled behaviour caused by the amazing boost in our intellectual proficiency. First of all, is the diagnosis correct? Can we really speak of an acceleration of scientific progress? Or are we confusing that with normal technological developments? Our response is a definitive "no". And that is what makes the next twenty-year period so fascinating, exciting and worrying.

But science is almost always a synonym for the power of certain communities of people over others, via extremely powerful networks which

© The Author(s) 2019
J.-H. Lorenzi and M. Berrebi, *Progress or Freedom*,
https://doi.org/10.1007/978-3-030-19594-6_5

jeopardise democracy—our way of life, and our private lives—at the same time. There is nothing condemning us to this, but everything today points to a takeover of applied science and revolutionary technologies by businesses which are gradually substituting themselves for states, politics and the collective will, which could eventually lead to new forms of slavery.

## 5.1    A Great Scientific Leap

Remember, the peak of eighteenth-century scientific thought was only the empirical translation of technological development. In a way, it is the very origin of the first industrial revolution, where simple artisans conceived—with no theoretical basis—exceptional machines which were going to transform the world. But science re-established its supremacy during the second industrial revolution, that of the second half of the nineteenth century, where on the contrary, it was the scholars who enabled the technological innovations of the era—chemistry, electricity, the car—to appear and to become the basis of a new model of growth. We are seeing exactly the same process today, restoring the role of driving force of a new world to fundamental sciences and their intersection with each other, particularly physics and biology. As Étienne Klein reminds us, "Paul Dirca, one of the founding fathers of quantum physics, did not suspect that the very abstract thoughts which led him to the discovery of anti-matter enabled the invention of the positron cameras in our modern hospitals". And in the same way, Albert Einstein certainly had no suspicion that: "the equations of his gravitational theory, general relativity, would one day have to be taken into account to make our everyday GPS work properly" (Klein 2008). This gives the twenty-first century an extraordinary destiny and certainly enables us to predict a new, stunning development of the world economy. But we are still just at the very beginning of this transformation. Sometimes we may feel like we have already entered this new phase of human history: but we are wrong.

We often define the notion of a "great scientific leap" by referring to the now-famous Moore's Law. It only interests us here as an illustration of the speed of scientific and technological progress. Noticing that the number of transistors in a circuit doubled every year, while costs had remained stable since 1959, Gordon Moore, co-founder of the Intel Company, stated in 1965 that this extraordinary increase would continue for a long time. Moore then revisited his prediction in 1975 by maintaining that the number of microprocessor transistors on a silicon chip would double every two years. Another version of Moore's Law mentions the doubling of capacities like power or processing speed every eighteen months. Beyond all the different versions of this law, the essential point for our consideration is really the status we give it. Whereas at the start it was no more than a prediction, it has gradually become an absolute truth by dint of being proven.[1] Except that today, the exact question is: Should we expect a slowdown, a continuation or an acceleration of Moore's Law?

Incidentally, the history of Moore's Laws acts as an illustration of technological development, since it is technology itself and its limitations which constrain the development of technology in the end, simply because we remain within the same scientific model. Yet thinking ahead to what the twenty-first century will bring highlights the existence of real revolutions in science which are going to change the existing paradigm.

The twenty-first century actually got off to a very strong start. We are discovering that the latest advances are only the first elements of a very profound transformation in our way of understanding and eventually transforming our world. This has given rise to claims that there will be more changes in science during the next fifty years than during the previous four hundred (Kelly 2006). Obviously, it is not just a matter of listening to our technological prophets. They are content to announce impressive developments in processing capacity, but these are mundane concepts. This is about astrophysics, nanotechnology and quantum physics and how their interaction will build a radically new scientific paradigm. But make no mistake; the twenty-first century will belong to biology. It is the field with the most researchers, new results, added economic value, ethical challenges and above all, the most knowledge to gain.

In the field of biology and health, the sequencing of the human genome allows us to catch a glimpse, as we have already mentioned, of the extraordinary hopes for personalised medicine and screening for diseases (*Nature*, 2016). With the proliferation of genome data, future medical discoveries will now depend on our capacity to process and analyse that data (O'Driscoll et al. 2013). For some biologists, this could constitute a turning point in traditional scientific methods, because it is no longer a matter of positing a hypothesis and then verifying it, but actually using mathematical tools and predictive models to deduce significant associations from it (Sagoff et al. 2016).[2] Another striking example is that the conclusions of the ENCODE programme, which enables entire genomes to be read, have lead some researchers to think that there is in fact no deterministic vision to genetic make-up. It seems that the latter is really the result of a random driving force controlled at the cellular level and working along the lines of natural selection (*Nature*, 2012). This amounts to saying that the explanation for life may not found in genes, but in the cell taken as a whole! In genetic engineering, we have also seen new editing techniques like CRISPR-Cas9, heralding a revolution in gene therapy. And finally, in biotechnology, the development of induced pluripotent stem cells (iPS or iPSCs) could be an alternative to the use of human embryonic stem cells and the ethical problems they pose.

But the early twenty-first century has also been one of discoveries in the field of physics. The confirmation, in 2012, of the existence of the Higgs boson particle, which had been postulated in 1964,[3] turned the world of particle physics upside down. The Higgs boson is considered the cornerstone of the standard model which corresponds to the paradigm describing all elementary particles and the forces acting on them. It allows scientists to explain why particles have mass. Above all, this discovery validates the standard model and could open the way to a new physics, "beyond the standard model". Let's not forget another fundamental discovery: "gravitational waves"! On 14 September 2015 at 11.50 a.m., two antennae from the American Ligo Observatory, located 3000 kilometres apart, allowed researchers to confirm Einstein's prediction. Both antennae observed a gravitational wave within around two-tenths of a second, or to put it another way, a vibration in space-time.

That particular day, there was a collision between two black holes: celestial bodies which are so compact, the intensity of their gravitational field prevents any type of matter or radiation from escaping them. The collision between these two black holes, respectively, 36 and 29 times the mass of the sun, produced strong cosmic waves, in the manner of a pebble thrown into a pond. The cosmos is actually described as curved and smooth and any mass moving inside it produces dents or hollows in space-time. Yet until now, it had been impossible to observe such waves in the universe. Moreover, and this is the whole irony of the thing, although Einstein accurately prophesied the existence of gravitational waves, he also believed humans would never be capable of detecting one. This discovery revolutionises the field of astrophysics by finally proving the existence of black holes and also opens the field to a new physics, heralding the beginning of an era where humans may be able to listen to the sound emitted at the moment of the Big Bang: the universe's first breath.

But the most intriguing idea is the predicted coming together of disciplines, where quantum mechanics could theoretically play a role in biology, even though these disciplines remain complete strangers to each other today. Before the twentieth century, crossovers between biology and physics were rare (Lambert et al. 2013). Biological systems were often considered too complex to be analysed via mathematical methods. How could a series of differential equations or physics principles elucidate a physical reality as complex as a living being? At the beginning of the twentieth century, with the arrival of more powerful techniques and microscopes, researchers began to explore the possibilities of physical and mathematical interpretations of microscopic biological systems. Progress in this field is fast at the moment and many branches of physics and mathematics have found applications in biology, from statistical methods used in bioinformatics to the mechanical properties observed in cells at the microscopic level.

Quantum mechanics can be applied to other fields, such as information technology. Originally the physicist Richard Feynman was one of the first, in the 1980s, to anticipate the different applications of the laws of quantum mechanics, particularly in information technology. Unlike a classic computer, which works on binary data to complete its calculations, a quantum computer will exploit the quantum properties of the subject. So it doesn't work on bits (0 or 1) but on qubits which have the unique property of having several values. What does this mean in practice? Quite simply that with quantum properties such as superposition[4] and entanglement[5] it becomes possible to carry out many more procedures on the data. A quantum computer of that kind would possess a colossal computational power and would be capable of massively parallel processing. Forget fears of an eventual slowing down of Moore's Law! Moreover, in this race towards the supercomputer, we could very well mention neurocomputers. These are computers in which the processor imitates the working of a group of brain cells, so the machine no longer executes a list of instructions, but makes millions of units interact, exactly like brain neurons. The most prestigious technology companies and some governments are not hiding their ambitions in this field of research, but the race towards the quantum computer seems to be little more intense.

We could also bring up nanotechnology as evidence of the key discoveries of the twenty-first century. Why can't we write the 24 volumes of the Encyclopaedia Britannica on the head of a pin (Feynman 1959)[6]? That was how Richard Feynman, in 1959, invited his audience to examine an aspect of physics which had not yet been explored—that of the infinitely small—during a speech given before the American Physical Society. Later, in 1974, Professor Norio Taniguchi (1974) used the term "nanotechnology" for the first time in order to introduce the manufacture of precision materials. But it is the American engineer Eric Drexler who takes back control of the term in 1979 and develops, in his book entitled *Engines of Creation: The Coming Era of Nanotechnology*, the first concepts of nanotechnology.

What are we referring to when we talk about nanotechnology? There is not really a clear definition which has an international consensus. But some define nanotechnology and nanoscience[7] as all the studies and processes for fabricating and manipulating structures, apparatus and material systems on an infinitely small scale: at the scale of nanometres,[8] the unit of measurement of atoms. For purists, the approach used in nanotechnology is known as "bottom up": it starts at the smallest and goes towards the biggest. In fact, one can understand the concept of the "bottom up approach" by saying it involves the opposite process to that of a sculptor with stone! As Michelangelo put it so well, sculpting starts with the biggest and ends with the smallest: "Every block of stone has a statue inside it and it is the task of the sculptor to discover it" (Tulevski 2016).[9] The approach used for nanotechnology is exactly the opposite method. We no longer start with a block of stone, but with a pile of dust and these millions of dust particles must be assembled to make a statue… Nanotechnology, then, is the manipulation of elements at an atomic scale in order to manufacture bigger components, which, first and foremost, have new physicochemical properties. Nanomaterials have the particular quality of possessing optical, electric and magnetic properties which are different from those they would have at a macrostructure level.

Investments in the nanotechnology sector are considerable and in a state of constant change throughout the world. Public investments, for example, multiplied sixfold in Europe between 1998 and 2003, and by eight in the United States and in Japan. From 2007 to 2013, the European Union allocated a budget of 3.5 billion Euros (French government, 2010)[10] in this sector. We must, of course, add private sector investments to this. It is estimated that worldwide there are more than 1,300 product families which incorporate nanomaterials (Nanotech project 2011),[11] whereas only 45 were listed in 2005. The American Federal Government research and development programme (NNCO 2016) actually encourages research into exploiting the convergence of nanotechnology, biotechnology, information technology and cognitive science—the famous NBIC Convergence—even though, for some researchers, talk of convergence remains debatable (Klein 2011).

What practical applications can we expect? They are very numerous and they cover all fields: energy, health, electronics, groceries, cosmetics, etc. But let's look at a few examples (Klein 2011). In the energy sector, using nanotechnology would allow us to vastly improve the performance of batteries, such as car batteries, and make them cheaper. We could also improve the yield from solar panels and photovoltaic cells; optimise hydrogen storage; or even manufacture materials which are as hard-wearing as steel and as light as plastic, which could lead to a reduction in the energy consumption of some industrial machines.

In the electronics sector, nanotubes could take over from transistors and enable the miniaturisation of components. Let's also take examples from the textile industry. Integrating nanocomponents into clothes could make them multifunctional. Why not imagine a fabric capable of harvesting the energy of the human body and transforming it into electricity, or dream up a fabric made from nanosensors which could identify the physiological condition of the wearer of a garment made from it? In the health sector, nanomedicine predicts the implementation of targeted treatments and nanodrugs which can target diseased cells. Target cells and organs can be reached specifically using nanovectors, which concentrate the medicinal molecules. Nanotechnology will make it possible to introduce miniature prosthetics in areas as complex as the human brain. Nanoparticles could even play a role in the clean-up and purification of water. Nanoparticles could be used as a filter capable of trapping arsenic, which poisons some world populations, for example, in Bangladesh. The applications of nanotechnology are so numerous that we are justified in foreseeing a radical change in consumers' way of life, which is a sign of a real step into the new industrial revolution.

All of these discoveries lead to a new approach to the philosophy of science, the consequences of which could lead to a real revolution in our view of the world. The term "technological singularity" is used to describe this disruption in current thinking. The technological singularity actually describes a predicted exponential curve in the evolution of knowledge: a true explosion of knowledge. It suggests that a scientific discovery will be able to shake up human civilisation in such a violent way that all the fundamentals of society will no longer be at all capable of being conceived with our current concepts, methods and values. The term "singularity" actually refers to the gravitational singularity, a region of space-time where some measures of gravitational field become infinite, making it completely unique, because current scientific knowledge would no longer apply in such a zone. And for a good number of experts, the technological singularity will actually arrive via artificial intelligence… Vernor Vinge, in an article in 1993 (Vinge 1993), already anticipated the beginnings of an artificial intelligence which will surpass human capabilities sometime around the 2030s. For Ray Kurzweil, the technological singularity will blossom in 2045 (Kurzweil 2006) when, he says, there will be progress in all areas, equivalent to 20,000 years of human evolution. From this, he theorised a law, which extends Moore's Law to all technology linked to nanotechnology, biotechnology, information technology and cognitive science. He calls it "The Law of Accelerating Returns". Its final outcome would mean almost unlimited power for humans over their brains and over matter.

Behind the concept of the technological singularity looms a worrying school of thought: Transhumanism, the movement which extols the use of technology in order to improve human physical and mental characteristics. We could describe it as potential new scientific utopia which wants to give human beings a kind of superhumanity. But as always, it is really about a power grab of people's minds and behaviour. The prime illustration of that power is asserting itself right now, through the growing influence of networks.

## 5.2   The Stranglehold of the Network Communities

People have organised themselves via network communities for a long time and the power that some believe they have is obviously not a new phenomenon. The novelty lies elsewhere, in the development that some networks have been able to achieve so quickly and on such a large scale, thanks to technology linked to the Internet. In order to illustrate this new stranglehold, let's look back over several worrying events which all took place during the 2010s. Of course we are describing the dark side of networks here, which obviously is in a tiny minority. However, it highlights the incredible, horrendous effectiveness of manipulation by network communities.

Let's begin with the so-called Islamic State or Daesh and their online recruitment process. For terrorist organisations, nothing beats social networks for conveying propaganda messages and videos in order to approach new recruits. According to the CPDSI (CPDSI 2014)[12] report, the propaganda process begins when a person stumbles across messages or videos while they are searching the Internet. Naturally, these videos consist of: conspiracy theories, a call to re-establish justice in this corrupt world and finally an appeal for partisan forces scattered around the world to wage "jihad". That is the moment when the gears are set in motion… Based on videos viewed, the recommendation algorithms will continue to supply the eventual supporter with a continuous flow of similar videos, containing the same propaganda messages, the same plots to be thwarted, the same threats to anyone who does not follow the precepts being presented. The productions are supposed to galvanise viewers, the video-montages to excite minds and the soundtracks incite awareness. The mission they are called to is essential; the videos encourage them to act! There then follows a phase of questioning all the values and beliefs the eventual supporter seems to have relied on until now. In order to continue the connection with that universe which claims to decode global conspiracies, the new supporter extends their experimentation to groups on social networks, all of which are manipulated by recruiters. The networks are global, with no borders, and in this way, they can gather tens of thousands of individuals, indeed far more than that, around a common cause. For the recruiters, all that remains is to sort through the grass roots to gain access to all sorts of profiles— men and women, young or less so—and to keep up the sales pitch that began with videos in order to select their next recruits. The figures are obviously not confirmed. In 2016, there was an estimate of between 46,000 and 90,000 Twitter accounts supporting the so-called Islamic State (Berger and Perez 2016), and between 500 and 2000 accounts which showed intense activity. Twitter incidentally took the initiative in 2015 to suspend 125,000 accounts for having "threatened or promoted terrorist acts". For its part, the Google search engine attempts to contain the radicalisation process by posting AdWords links for certain extremist searches which redirect users towards de-radicalisation sites. The efforts made by the technology companies to limit the influence of terrorist social networks are beginning to develop, and there was a lively debate around the famous "back doors" designed to make certain data accessible to states. But it is hard not to be devastated when an antiterrorist organisation[13] announces in 2016 that Daesh may have developed its own secure messaging system.

The influence of the network communities described here—still focussing on the dark side, to demonstrate the problems they can pose—can also be gauged via the scale of the impact of fake news, the mass circulation of false information in order to influence public opinion. The years 2016 and 2017 were punctuated by major political events all over the world. For some, such as the United States, France, the Netherlands or Germany, it involved presidential elections. For others, it was referendums. And faced with such high stakes, networks can sometimes play a not insignificant role in influencing some undecided voters. Do you remember the voices of protest against certain platforms, such as Facebook, for example, who were accused of having allowed articles containing disinformation to proliferate and of playing a role in the United States election? Or these surnames: "Ali Juppé" and "Farid Fillon", widely shared on some networks in France accusing certain candidates of a supposed collusion with Islamic organisations? And what can be said about the campaign of disinformation which was led in order to encourage votes one way or the other in the UK? But we can wonder, with some justification, about the influence of these networks. Can election results be influenced to this extent by false information, shared on a mass scale on networks? For Robert Epstein,[14] the answer is clear. Yes, articles can influence undecided voters (Epstein and Robertson 2015). Martin Moore[15] gives the same answer, since such articles, he says, are "bound to have an impact on your point of view". According to Epstein, influencing undecided voters already happened in India in 2014. Back in 1876, Rutherford B. Hayes got himself elected nineteenth president of the United States with the help of the telegraph company, Western Union, which delivered him copies of his competitors' telegrams and manipulated the newspapers in his favour. Today, according to Epstein, it would seem that you need to be chosen by the Google search engine in order to win elections. For him, the networks represent "the most powerful mind-control machine ever invented in the history of the human race" (Cadwalladr 2016).[16] Moreover, private companies already deploy significant resources in an attempt to take control of their online reputations, with teams entirely dedicated to surveying social networks. For the political classes, it is not a matter of taking care of their reputations, but of being able to prevent the networks from turning into platforms for the mass spreading of false information. Now it is certain that some technology companies are announcing counter-attack measures deploying new tools, but it is still very early to confirm whether such tools will be effective in the struggle against the campaigns of disinformation.

The investigation by Guardian journalist Carole Cadwalladr reveals another type of influence through the networks, but more subtle than simply spreading false information. All of the power and intelligence of search engine algorithms consist in exploiting users' searches, so as to be capable of recommending or predicting the end of key words to type when we want to search the Internet. The examples given by Carole Cadwalladr are terrifying: a mixture of anti-religion, racism and sexism. Why does an algorithm offer such suggestions? Does it mean that the mass of users of the search engine associates such keywords with such searches, a little in the manner of the "invisible hand", but applied to data? The explanation given is confusing: "Search predictions are generated by an algorithm automatically, without human involvement. The algorithm is: based on several factors, like how often others have searched for a term; designed to show the range of information on the web."[17] For Danny Sullivan,[18] algorithms actually favour popular results rather than those which carry authority, because, he says, it enables technology companies to be more lucrative (Ertzscheid 2016).

From a list of more than 300 websites relaying false information, Jonathan Albright[19] mapped the connections and links generated by platforms. The results are striking: he counts more than 23,000 pages and 1.3 links generated. It is a vast canvas which looks like a virus when it is visualised in 3D, which represents a network of distribution, a whole ecosystem, favourable to misinformation, which stifles information relayed by traditional media. What a paradox! On the one hand, there is the will to encourage criticism and the plurality of the media. On the other, we are witnessing the emergence of a propaganda machine which has become out of control, as Charlie Beckett[20] stresses.

But the stranglehold of the networks does not limit itself to influencing just human beings. Machines can also experience it and come under the influence of the networks. Let's take the example of a chatbot which was pirated in the space a few hours in March 2016. As a reminder, a chatbot is an artificial intelligence programme which allows a message service to converse with humans. First, the artificial intelligence had to learn from contact with humans and to embody "an American teenager" capable of chatting online with young "Americans aged 18 to 24". The programme was going to be able to learn in real time: the more you chatted with the artificial message service, the more intelligent it became. Faced with that promise, several users agreed on forums to organise themselves to coordinate an operation to pirate the learning process. The result was that in barely 24 hours, the message service in question had become racist, because the pirate operation had succeeded perfectly.

For Rebecca MacKinnon,[21] there is an urgency: we must react. People would be unlikely to question the content available on the Internet. They consider the Internet like "the air that we breathe and the water that we drink" (Cadwalladr 2016).[22] According to a study (Epstein 2016), almost 63% of people in the world would place more trust in search engines than in the traditional media. In this way, these network communities represent a power and strength of intervention which has not been equalled to date: in most cases for the best, but sometimes for the worst too.

## 5.3   Widespread Intrusion

Digital technology has brought us knowledge and access to sources of learning which were previously unknown. But like anything man-made, they can enable completely unacceptable situations to occur. In this case, we are referring to intrusion into our personal lives to the extent that it looks like a real takeover. This drive for surveillance is nothing new. It is one of the constants of all political and economic powers to different degrees; which brings us to Jeremy Bentham.

At the end of the eighteenth century, Bentham presented his Panopticon to the French National Assembly. This was a means of surveillance which could be applied to all institutions—prisons, hospitals, schools, etc.—essentially based on a building design which enables the maximum number of people to be watched in the most economical way. In a panoptic prison, the building is circular and comprised of six storeys. The cells have grilles which open on to an interior courtyard. At the centre of this circular building, there is a central tower where the guards circle around so that they can survey all of the prisoners. The principle is quite a simple one after all: it consists of seeing without being seen, or rather it consists of giving the observed people the feeling that they are constantly watched and thus controlling their minds by discouraging any deviation in their behaviour. In order to understand the reasons for this invention by Jeremy Bentham, who is known as the precursor of liberalism, let's go back to the underlying principles of Bentham's philosophy. He considers that people only comprehend their own interests via the relationship between pleasure and pain and that they constantly seek to maximise their own interests and personal pleasure. That is why, according to Bentham, if a state wishes to govern well, it must absolutely seek to maximise the pleasure of the greatest number and thus restrain people from being tempted to act in a way which is contrary to the interests of the community. For Bentham, it was impossible to trust humans, those "potential delinquents" guided only by their own interests. They absolutely must be kept under surveillance. The Panopticon is the solution to the problem of channelling and mastering people's temptations and giving them all the feeling of being watched.

In fact with Bentham, when we talk about surveillance, we are referring to a general transparency, since each individual must be watched by all. He thus evokes the term "panopticism" and so quite close connections appear between the notions of happiness, liberty, surveillance, security and transparency. As Christian Laval reminds us: "we can only be in such a context [of security] when we think that the capacity of others to inflict harm is limited by fear of punishment and thus by the probability of paying dearly for wrongdoings or crimes. We can therefore act even more freely when we know that threatening people are under surveillance and control. The more we want freedom, the more we want to feel safe, and the safer we want to feel, the more we feel the need to be under surveillance and the more surveillance and control we demand" (Laval 2012). The Panopticon is not, in the end, the exclusive property of totalitarian regimes. It is, according to Bentham, equally the prerogative of a democratic society.

Can we use the image of the Panopticon to describe early twenty-first-century society? And if so, who is involved: which countries and which businesses? Video surveillance cameras have become common currency in cities. In France, there could be 935,000 (CNIL 2012)[23]; in the UK, there are definitely more than 4 million (Giannoulopoulos 2010). Some are even intelligent, biometric cameras, capable of facial recognition. We all know, but choose to forget, that individuals' movements can be tracked via several different channels. The expression "surveillance society", taken from the philosophical works of Michel Foucault (1975), is now a common expression in Anglo-Saxon countries, to the extent that it is used in the official reports of the House of Lords and Scotland Yard (Laval 2012). But as Christian Laval reminds us, if we go back to the definition of a surveillance society used by some British researchers, it corresponds to a society which uses technical and technological means to follow, record and store individuals' movements, data and activities. The creation of the database is a means of preventing danger, anticipating risks of criminality or attempting to reduce them. The digital version of Bentham's panopticism is described as horizontal panopticism, or inverse panopticism, in the sense that we now have surveillance of all by all on the Internet. David Lyon (2007) states that by surveillance, we should understand a systematic, routine and targeted attention to the behaviour of individuals for the purposes of domination, influence, power and/or protection. But what is more interesting for us to analyse is the fact that modern society is not satisfied with just one layer of surveillance. Unlike its predecessors, it is establishing complex, multi-layered systems.

First, there is a new form of surveillance or self-testing, which relies on self-tracking tools which measure personal performance and other indicators. Then, and this is what interests us, there is horizontal, or inverse surveillance, routed via social networks, through which everyone can spy on everyone, all the time, everywhere. On social networks, everyone can develop and manage their digital identity, display their work, inform others about their competences and convey their connections. But most importantly, everyone is permanently visible! For Dominique Piotet, "disappearing has become impossible: the entanglement between our physical life and our virtual life has become so enmeshed that it seems impossible today to completely separate them" (Piotet 2011). Especially since the enthusiasm for social networks encourages a lot of users to indulge in confessions, waiving anonymity. But in the world of horizontal surveillance, unlike the vertical Panopticon, there is no appetite to punish or to anticipate the deviations of "potential delinquents". It is more about a quest for visibility and popularity. As Miyase Christensen (2010) says: "there are many of us taking part in this new type of surveillance on a voluntary basis, often without being aware of its scope". Finally, for Simon Borel, the reticular individual can become "His or her own despot, when they hand over, on an almost voluntary basis, enormous amounts of personal data, which are used by states and by the large digital companies" (Borel 2016).

What can we also say about those technology companies who keep their users' data, by using cookies and other virtual informants? The virtual society claims to be contributory and participative. Users cooperate and interact with private companies, they give their opinions, ranking some of them and sharing their preferences. But at the same time, this is a new form of market monitoring, with technology companies able to exploit the mass of information collected. By offering free, informative or cultural products, some websites obtain the consent, indeed the obsessive interest, of the consumer, so that they can harvest this manna of information and then be in a position to process it. They process it in order to get to know the consumer's behaviour better, to better anticipate their needs or direct their choices, by suggesting articles which are most likely to interest them. Or they may process the information in order to monetise it with targeted marketing and advertising sales.

And finally, as part of this widespread intrusion, there is intrusion by the state. This is certainly the most basic and oldest form of surveillance in history. Many commentators believed the Internet and digital tools were actually going to represent forces for the liberation of nations from all tyrannies in the early twenty-first century. With these communication tools, forget dictatorships, forget totalitarian regimes! The Internet was going to allow nations to express themselves at last: to convey revolutionary ideas on a massive scale in a virtual space and to imagine new forms of freedom. But this angelic, naïve vision of the messianic role of the Internet is obviously not a shared one. For Evgeny Morozov (2011), it is nothing of the sort. The Internet does not play the liberating role attributed to it. In his view, authoritarian regimes have succeeded in drawing lessons from anti-establishment demonstrations—e.g., in Tunisia or in Egypt—and now they too know how to exploit Internet networks in order to monitor their populations. Let's examine some of Morozov's examples.

After the demonstrations of 2009, the government in Iran launched measures to monitor the population more closely. According to Morozov (Millot 2011),[24] the day after the demonstrations, the Iranian authorities collected photos of the demonstrators and published them on official websites to encourage people to mobilise and identify the protestors. Monitoring techniques in Sudan are equally accomplished. Morozov says that the secret services there launched false demonstrations on the Internet in order to arrest protestors who gathered to demonstrate in the street. Other countries, such as Russia or China, prefer to use propaganda rather than censorship, because it seems more effective to them. "If a Chinese blogger accuses a local councillor of corruption, rather than removing his post, the authorities can activate pro-government bloggers: they will discredit him or her and insinuate that this protestor is acting on behalf of the CIA, the West or Mossad". The sociologist Gary King sees in the Chinese government's deployment of all these means of control on the Internet the "most important effort ever made to selectively censor human expression" (Kauffmann 2012).[25] Morozov points out the liberating utopia the Internet can constitute for totalitarian regimes; but when it comes to implementing these forms of surveillance, Western countries are not innocent either.

Remember Edward Snowden? With the help of two journalists, the young engineer divulges a considerable volume of documents which prove that the United States practises massive Internet surveillance on a worldwide scale. He reveals the different information-gathering programmes[26] online. The world then learns that the NSA[27] has access to data hosted by the large digital companies. This is a precise example of vertical surveillance, by a state, which actually relies on horizontal surveillance on the Internet and social media.

Among all of this widespread intrusion, it is perhaps the disconnect between the shockwaves provoked by the revelation of the Snowden affair and the indifference of the overwhelming majority of users which is most surprising. Snowden claimed to be a freedom fighter. He wanted to give "a chance to society to decide if it wants to change". It is true that subsequently, technology companies opposed the installation of back doors—hidden portals—allowing the NSA to access their data. Messaging platforms and social networks have implemented encryption systems… But for the overwhelming majority of users on the Internet, have we observed, following all these revelations, any kind of change in Internet use behaviour? Should we believe Bernard Harcourt (2015) that the population has given its consent to be surveyed and that we are so fulfilled by the enjoyment of the digital services we are offered that we cannot do without them? This just shows the intrusion is widespread, involves states and individuals, public and private spheres, by means of technologies which never stop improving. What system of regulation would allow us to limit or eradicate the inevitable abuses? That is the entire question of how twenty-first-century society functions.

## Notes

1. Between 1971 and 2001, the concentration of transistors doubled every 1.96 years.
2. Mark Sagoff. Data Deluge and the Human Microbiome Project, vol. XXVIII, no. 4, Summer 2013; Larry Hardesty. Making Big Data Manageable, MIT News Office, December 14, 2016; Alison Abbot Pharmaceutical Futures a Fiendish Puzzle. *Nature*, no. 455, October 29, 2008, pp. 1164–1167.
3. Independently by Robert Brout, François Englert, Peter Higgs, Carl Richard Hagen, Gerald Guralnik and Thomas Kibble.
4. Having several values for a certain observable quantity.
5. When two objects, although separated in space, cannot be described separately.
6 Feynman R (1959) From a speech entitled: There's Plenty of Room at the bottom, December 29, 1959.

7. The CNRS (French National Centre for Scientific Research) distinguishes between nanoscience, which seeks to formulate laws and study nano-objects in the laboratory and nanotechnology which aims to create marketable products.

8. 1 nanometre = 1 billionth of a metre.

9. Example used by George Tulevski. The Next Step in Technology, TED Conference, 2016. Available via https://www.ted.com/talks/george_tulevski_the_next_step_in_nanotechnology?language=en.

10. Rapport sur les nanotechnologies du groupe de travail du Conseil national de la consomation, 2010 (Report on nanotechnology by the working group of the French National Council on Consumption). Available via https://www.economie.gouv.fr/files/directions_services/cnc/avis/2010/140610rapport_nanotechnologies.pdf.

11. The project on Emerging nanotechnologies, 2011. Available via http://www.nanotechproject.org/.

12. Centre de prevention contre les dérives sectaires liées à l'islam (Centre for prevention of cults linked to Islam). Available via http://www.cpdsi.fr/2014/11/.

13. Ghost Security Group: an antiterrorist hackers' collective.

14. Senior Research Psychologist at the American Institute for Behavioural Research and Technology in Vista, California.

15. Director of the Centre for the Study of Media, Communication and Power at King's College, London.

16. Quoted in Cadwalladr, C (2016) Google, Democracy and the Truth About Internet Search. *The Guardian*.

17. Google search help page, available via https://support.google.com/websearch/answer/106230?co=GENIE.Platform%3DAndroid&hl=en.

18. Founder of SearchEngineLand.com.

19. Assistant professor of communications, Elon University, North Carolina.

20. Professor in the department of media and communications, London School of Economics.

21. Director of the Ranking Digital Rights project at the New America Foundation.

22. Quoted in Cadwalladr, C (2016) Google, Democracy and the Truth About Internet Search. *The Guardian*.

23. CNIL (French national commission for information technology and freedom) (2012) Vidéosurveillance/vidéoprotection: les bonnes pratiques pour des systems plus respectueux de la vie privée. (Video

surveillance/videoprotection: good practices for systems which are more respectful of private life.)

24. Interview with Lorraine Millot: Le Net, instrument de liberation et d'oppression (The Net: instrument of liberation and oppression). *Liberation*, March 5, 2011.

25. Reported by Sylvie Kauffmann. Weibo, la "terreur Internet" des dirigeants chinois (Weibo, the "terror internet" of the Chinese Leaders). Le Monde, September 11, 2012.

26. PRISM, XKeyscore, GENIE.

27. US National Security Agency.

# References

Berger, J. M., & Perez, H. (2016, February). *The Islamic State's Diminishing Returns on Twitter*. George Washington University. Available via https://cchs.gwu.edu/sites/g/files/zaxdzs2371/f/downloads/Berger_Occasional%20Paper.pdf.

Borel, S. (2016). Le panoptisme horizontal ou le panoptisme inversé. *Tic & société, 10*(1), 2016.

Christensen, M. (2010). *Facebook Is Watching You*. Manière de voir, no. 109, mars 2010, pp. 53–55. Available via https://www.monde-diplomatique.fr/mav/109/CHRISTENSEN/19307.

Epstein, A. (2016). *People Trust Google for Their News More Than the Actual News*. Quartz. Available via https://qz.com/596956/people-trust-google-for-their-news-more-than-the-actual-news/.

Epstein, R., & Robertson, R. E. (2015, August). The Search Engine Manipulation Effect (SEME) and Its Possible Impact on the Outcomes of Elections. *Proceedings of the National Academy of Sciences of the USA (PNAS), 112*(33). Available via https://www.pnas.org/content/112/33/E4512/tab-article-info.

Ertzscheid, O. (2016). Quand on demande à Google si l'Holocauste a bien eu lieu… [When You Ask Google If the Holocaust Really Happened…]. *Le Nouvel Observateur*.

Foucault, M. (1975). *Discipline and Punish*. New York: Random House.

Giannoulopoulos, Dr., D. (2010). La vidéosurveillance au Royaume-Uni, la camera omniprésente: signe d'une evolution vers une "société de surveillance"? [Video Surveillance in the United Kingdom, the Omnipresent

Camera: The Sign of an Evolution Towards a Surveillance Society?] *Archives de politique criminelle*, 1/2010, no. 32, 245–267. Available via https://www.cairn.info/revue-archives-de-politique-criminelle-2010-1-page-245.htm?contenu=article.

Harcourt, B. E. (2015). *Exposed—Desire and Disobedience in the Digital Age*. Cambridge: Harvard University Press.

An Integrated Encyclopedia of DNA Elements in the Human Genome. *Nature, 489*(7414), 57–74. September 6, 2012. Available via https://www.nature.com/articles/nature11247.

Kelly, K. (2006, June 4). *Speculations on the Futures of Science*. Edge.org. Available via https://www.edge.org/conversation/kevin_kelly-speculations-on-the-future-of-science.

Klein, E. (2008). L'avenir de la recherché scientifique. *Études, 408*(6), 725–728. Available via https://www.cairn.info/revue-etudes-2008-6-page-725.htm.

Klein, E. (2011). *Le Small Bang des nanotechnologies*. Paris: Odile Jacob.

Kurzweil, R. (2006). *The Singularity Is Near*. London: Penguin.

Lambert, N., et al. (2013). Quantum Biology. *Nature Physics, 9,* 10–18. Available via https://www.nature.com/articles/nphys2474.

Laval, C. (2012). Surveiller et prévenir. La nouvelle société panoptique [Discipline and Prevent: The New Panopticon Society]. *Revue du MAUSS*, 2/2012, no. 40, 47–72. English version available via https://www.cairn-int.info/article-E_RDM_040_0047–discipline-and-prevent-the-new.htm.

Lyon, D. (2007). *Surveillance Studies: An Overview*. Cambridge: Polity Press.

Morozov, E. (2011). *The Net Delusion: The Dark Side of Internet Freedom*. Philadelphia: Public Affairs.

National Nanotechnology Coordination Office. (2016, October). *National Nanotechnology Initiative, Strategic Plan*. Available via https://www.nano.gov/sites/default/files/pub_resource/2016-nni-strategic-plan.pdf.

O'Driscoll, A., et al. (2013, October). Big Data, Hadoop and Cloud Computing in Genomics. *Journal of Biomedical Informatics, 46*(5), 774–781. Available via https://www.ncbi.nlm.nih.gov/pubmed/23872175.

Piotet, D. (2011). Comment les réseaux sociaux changent-ils notre vie [How Social Networks Change Our Lives]. *Esprit* no. 376 juillet, 82–95. Available via https://www.cairn.info/revue-esprit-2011-7-page-82.htm?contenu=resume.

The Power of Big Data Must Be Harnessed for Medical Progress. *Nature*, November 23, 2016. Available via https://www.nature.com/news/the-power-of-big-data-must-be-harnessed-for-medical-progress-1.21026.

Taniguchi, N. (1974). On the Basic Concept of Nanotechnology. In *Proceedings of the International Conference on Production Engineering, Tokyo, Part II*. Japan Society of Precision Engineering.

Vinge, V. (1993). *The Coming Technological Singularity: How to Survive in the Post-Human Era*. San Diego, USA: NASA.

# 6

## A Disengaged Society?

There is a kind of tide of nostalgia, a kind of disengagement by society, both from understanding the world which is slowly appearing and from building a plan for the future. Society is as distracted by the small advantages technology brings as it is unaware of the extraordinary high stakes which the scientific revolution is bringing to fruition, particularly in genetics. Faced with sociological upheavals unknown since the second industrial revolution, society is sustaining two opposing visions of the coming months and years. One is rigid and closed: populism. The other is committed to redefining a modern humanism which is respectful to humans, their liberty and their freedom. It is here that our techno-prophets, our sorcerers' apprentices from all sides of the scientific revolution, can play a harmful role. While some scientists bring genuine intellectual disruptions to the human community, there are technologists of all types who are seizing on that knowledge to portray, without the slightest restraint, the world they have dreamed up, which is obviously to their benefit. But that will not last forever. It is within the realms of possibility that the architects of genome sequencing will soon offer solutions to improve the human condition and in doing so flout the most basic of our ethical rules.

© The Author(s) 2019
J.-H. Lorenzi and M. Berrebi, *Progress or Freedom*,
https://doi.org/10.1007/978-3-030-19594-6_6

This would just be a simple recurrence of the ever-present grasping for power by some group or other, if it were not for the fact that the way our society operates is weakening political discourse in a dangerous way and is suddenly giving overexposed power to these prophets. Never, since the nineteenth century, have human societies, whether in developed or emerging countries, been so divided, nor the labour markets so polarised, and the middle classes produced by Fordism so fragile. The political consequences are not slow to appear: they are translating into radicalisation and populist movements which express so well the feelings of loss of status of a majority of the citizens of developed countries. For the first time in a long time, the spread of technology is playing a part in the gradual disappearance of politics and the emergence of a rejection of our democratic society.

## 6.1    A Multi-speed World

This disengagement is not everyone's concern, but it is most people's. Globalisation has not kept its promises. Because of its excesses, it has itself engendered threats and fears which make it crucial to the success of those who wage campaigns for past glories and for regaining control of national destiny. Disengaging when faced with the future, abandoning progress, as Jean Pisani-Ferry (2017) maintains, is the greatest danger of the early twenty-first century. Nostalgia is our contemporaries' mindset and reactionarism their credo, Pisani-Ferry says. The whole of the west and many other regions of the world seem to be swept up in the storm of illiberalism.

He reminds us that the causes of this are known. Progress has, in a way, betrayed the West. This is proved by the recession in which young thirtysomethings live, the gravity of the financial crisis and dropping productivity rates. It is also proved by: the extreme polarisation of the labour market, the increasingly unequal redistribution of incomes and the geographical segregation of the worst-off and the exclusive set who are on first-name terms with globalisation. Progress does not have a good press anymore, because it only seems to benefit certain people. This is the worst refutation of the promises made by the post-war social advances.

This is so true that everyone seems to be addressing this subject now. And lo and behold, it was even the subject of a forum meeting at Davos 2017. There are obviously extremely simple reasons behind this idea of abandoning all hope in progress: mainly, all types of inequalities. That is what a multi-speed world is.

This expression originated with Oxfam, the British NGO which specialises in inequality. According to Oxfam, in 2017, the eight richest people earn more than the poorest half of the world's population (Oxfam 2017). The expression two-speed world is an attempt to express the immense discrepancy, the like of which has not been seen for a century, between citizens of the same society.

And yet this situation plunges experts and analysts into profound perplexity. The list of authors (called "the religion of inequality" by Richard H. Tawney [1964]) for whom inequality constitutes a fact of life, or a good thing in itself, is very long. However, it is possible to find two fundamental principles in it: the belief that property is natural and that individuals are de facto unequal. Clearly, these two pillars of order in every society often appear linked, except perhaps for John Stuart Mill, who considered private property as one solution among others and who went as far as to envisage limiting the right of inheritance in order to get closer to the principle of "equal point of departure"[1] he was the father of this vision, which broke with believers in natural inequality.

But Stiglitz would go much further. According to him (Stiglitz 2015), inequality does not just pave the way to reactionarism: "it kills". He builds on research by Angus Deaton and Ann Case which reveals "declining life expectancy and health for middle-aged white Americans, especially those with a high school level of education or lower. Among the causes were suicide, drugs and alcoholism". Yet, still according to the Nobel laureate, this study, which dismantles the very widespread racial argument, shows that the United States is a society which is increasingly divided between the richest 1% and all the rest. Disparities in income and life expectancy are fellow travellers in a country where the median income of a salaried worker is lower than it was forty years ago. He concludes, having never considered GDP a good indicator of well-being, that "the American middle class – the very quintessence of what a middle class is – is on the way to becoming the first ex-middle class on the planet" (Stiglitz 2015).

All of this could make the richest 1%, or indeed 0.1%, the only winners of globalisation and its crises. Chrystia Freeland (2013) points out that this process is just as active in communist China, in Canada or "in cosy social democracies like Sweden, Finland and Germany". She also highlights the fact that Warren Buffet shows himself to be an astute observer of this explosive concentration of wealth in the hands of a few. Whereas "crony capitalism" can explain this phenomenon in part, it may be corruption or globalisation coupled with the technological revolution which has allowed certain very gifted minds to become very rich, quickly. Chrystia Freeland also refers to what she calls "the global educational arms race". This new class wants the best for its children: to establish a lineage, a sort of new aristocracy. Increasing the numbers of the daughters and sons of the 1% are shining at Harvard, Stanford or MIT, underlining the relevance of the "Great Gatsby Curve" described by the economists Alan Krueger and Miles Corak: as social inequalities widen, social mobility decreases.

Which brings us back to Oxfam. The NGO denounces "an economy for the 1%"[2], with figures to prove it. This exponential growth in economic inequality is known to be equally undermining to growth and social cohesion. This danger has been taken very seriously for several years by all international organisations, who have made it one of their key issues alongside terrorism, climate change, pandemics and economic stagnation. And while reducing inequality now seems to be an almost insurmountable difficulty in practice, each institution's statement is more alarming than the next's. The OECD is concerned, for example, about the "surge" in incomes benefiting the richest of the rich (OECD 2014).

Following the IMF and the International Labour Organisation (ILO), it falls to the World Bank (2016) to go beyond its brief and engage with inequalities within rich countries themselves and with the new "white people's poverty". As for Stiglitz, he condemns another aspect of these dysfunctions in our society: the fact that the inequalities actually hit younger generations much harder. The social ladder will not work for them because "job insecurity follows them throughout their lives" (Stiglitz 2016, March). He states that any reforms have actually reinforced the existing system, which has engendered feelings of social injustice and defiance in young people towards the "elites": in this case the political and financial establishment. These inequalities between generations, which unfortunately get less attention than others, carry with them the seeds of a conflict which we are struggling to outline, because history has given us so little experience of it.

How can we link this explosion in inequality with the explosion in technology? The ultimate justification of the contrary nature of our age is the fact that everyone needs technology. Technology is basically the religion of the beginning of this century, with a fundamental difference: paradise is not for all, in the afterlife; it is for a few, right now, on Earth. In France, our generation has often wondered what life was like under the *Ancien Régime*,[3] in a society with such clearly defined classes: the aristocracy, the clergy and the third estate. But there is nothing new under the sun. Justifications, divisions, social classes simply change their appearance, but not their essence. This so-called era of the 1% cannot fail to remind us of the *Ancien Régime* society as Marivaux described it in *La Vie de Marianne* (Marivaux 2007). It paints a very nuanced portrait of the peasants or common people, the artisans, the servants, the petit bourgeois, the minor aristocracy, members of fashionable society, arrogant noblemen, the devout, nuns and of course the whole ecclesiastical hierarchy.

It reveals the hypocrisy and conformism of a hierarchical society with fixed rules: a society enclosed by prejudices, and, above all, where Marianne has no place. She is told: "You may be young, distracted, reckless, anything you please; but you may not forget your station, when it is as sad, as deplorable as yours […] you are an orphan, and an orphan whom nobody knows, who belongs to no one on the face of the Earth, who has no one to worry about them, forever unknown by your family, whom you also do not know, with no parents, no goods and no friends".

Transpose all of that to today's world. There are the 1%: the noble lords; the third estate: in this case, the middle classes; and then there are the bearers of God's word, which today means the technology prophets and everyone who obeys the instructions of these powerful leaders and repeats over and over again that the world will obviously be fashioned by their technology, giving a pseudo-scientific justification for the explosion of this multi-speed society.

## 6.2    The Shock to the Middle Classes

The concept of "the middle class" attempts to characterise a category of people who are neither rich nor poor. In fact, an official definition of the middle class does not exist. Some approach it in terms of income. In that instance, a middle class can be defined as the group of individuals whose income, after social benefits and before taxes, is between two-thirds of and two times the average income.[4] Another method consists of dividing the population into tenths and considering the middle class as the half of the population which is neither in the poorest 30% nor the richest 20% (L'Observatoire des inégalités 2018). Others, such as INSEE (the French National Institute of Statistics and Economic Studies) prefer to approach it in terms of standard of living. Yet others are focussing on socio-professional categories, or income gaps, or even on self-evaluation and feelings of belonging. It all remains quite fluid.

The lament for the inevitable disappearance of the middle class is already old. From the end of the 1990s, there were concerns about what was called "tearing the social fabric" (Lipietz 1998). Many others followed: sociologists, economists, journalists… In 2014, two books announced a planned demise. For David Boyle, this disappearance was the consequence of property speculation, which stops the middle class from having access to the thing that defines it: property.

Tyler Cowen states that the American middle class cannot survive the ongoing polarisation of the labour market (Cowen 2013), the digital revolution and the introduction of intelligent machines. He says 90% of that population is condemned to live in a state of "stupefaction", which we prefer to call "disengagement".

If Cowen and Piketty have had such great success in the United States, it is because they express loud and clear the long-held feelings of the American middle class who have been touched by a kind of despair since the 1970s. In a study by the American think tank the Pew Research Centre, the authors reveal that the blow to the American middle classes is much more violent than in other countries, such as France, for example. In 2012, the middle classes represented 67.4% of the French adult population, and only one adult in two in the United States.

This actually relates to a much older situation. In 1971, the American middle class actually represented just 60.8% of the adult American population, which was already lower than the percentage in France in 2012.

We can therefore understand why the subject of the decline of the middle classes is so present and so sensitive in public opinion. For more than forty years, the American middle classes have been haunted by the spectre of a level of poverty they thought belonged only to the Great Depression, and by the prospect of an apparently inevitable loss of status. It should be stated that their median income, in strong decline since 2008, is struggling to regain its 1996 level. The French middle classes, in comparison, have not suffered anything like such a shock, since their incomes have risen by almost 20% in that period, with a minor hiatus between 2009 and 2012. This obviously does not mean that they are shielded from such a development, simply that their timeline is very clearly lagging behind the United States.

The whole world, with the IMF at its head, is concerned about the gradual decline of the middle classes in developed countries, incidentally more for political than for economic reasons. But the analysis becomes more interesting when we explore generational issues.

Rather than studying the growing gap between different sections of the population, McKinsey carried out a study (McKinsey 2016) which took as a starting point the stagnation or decline in household income and examined whether successive generations were living better than their predecessors. The McKinsey results are striking! For 65–70% of households in the 25 developed countries studied, incomes—earnings and capital income—had stagnated or dropped in 2014 compared with 2005. That statistic is important because, between 1993 and 2005, this had only been the case for less than 2% of households. In absolute values, the results are as follows: whereas there were only 10 million people affected by that income stagnation period of 1993–2005, there are now between 540 and 580 million people affected over the period 2005–2014. Now of course, that does not mean salaries have necessarily dropped for two-thirds of households. But it does however call into question the assumption that future generations will live better than their parents and their grandparents… In short, today's young people risk being poorer than their parents. This is perhaps the end of the middle-class dream: a betrayal of the founding social contract of our modern society.

Let's go back then to the data from France Stratégie on the distribution of the population according to income class in France and the United States. We have carried out a simple projection to 2040, assuming the continuation of the trend observed between 1996 and 2012. This enables us to appreciate the changes undergone by the middle class and the way it has been gradually squeezed over time: very gradually in France and very quickly in the United States. But as we have seen, this was already a part of the last twenty years of America's history, as a country hit more heavily by inequality than others (Graphs 6.1 and 6.2).

We can also highlight the fact that states played a decisive role in maintaining social equilibrium during and after the 2008 crisis. The American government intervention was a lot less significant than the very generous interventions of other countries, particularly Sweden. Whereas market income fell for 81% of American households between 2005 and the end of 2013, it only stagnated or fell for 20% of Swedish households.

Such developments had significant repercussions, not just on household consumer demand and on GDP growth, but perhaps more so on popular opinion, particularly among the middle classes, whose discontent and loss of confidence in existing political and economic structures is measured year after year. The McKinsey Global Institute is very explicit on this subject. It says that whether it is in the UK or the United States, almost a third of those whose incomes are stagnating and who have no hope of any improvement are much more critical about international trade, globalisation and immigration than others. Recent political elections illustrate this perfectly.

In recent years, we have focussed a lot on income inequality in developed countries. This may have been a mistake. Stagnation and the noticeable decline in the incomes of the majority are much more significant. What is at stake is the threat of a slow, irreversible erosion of the middle classes who give structure to our society.

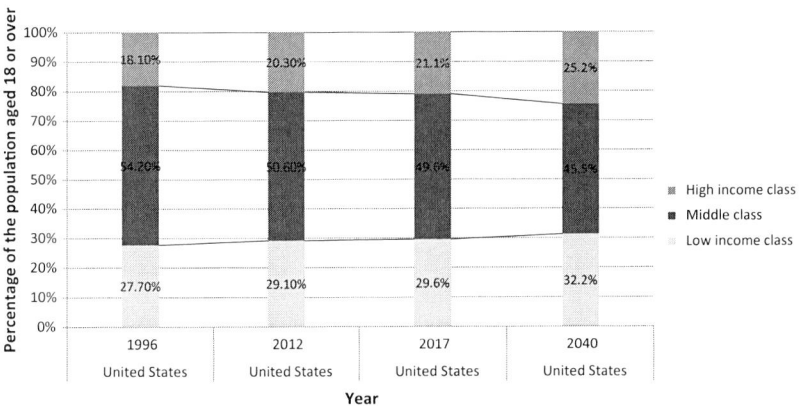

**Graph 6.1**   United States: Population distribution according to income class (*Sources* France Stratégie, ERFS Study by INSEE, Pew Research Centre and the authors)

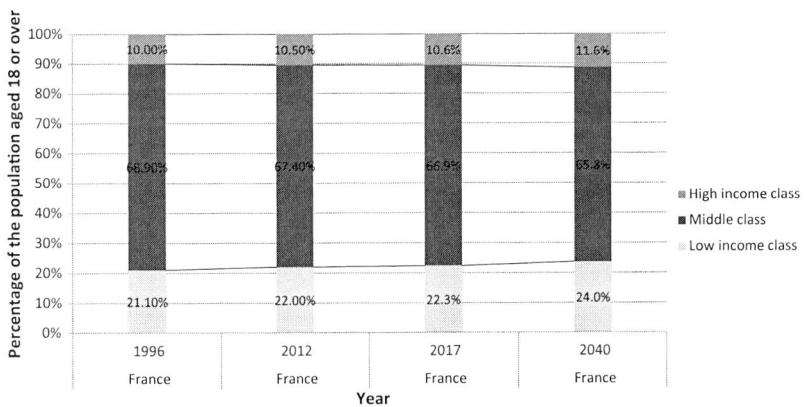

**Graph 6.2**   France: Population distribution according to income class (*Sources* France Stratégie, ERFS Study by INSEE, Pew Research Centre and the authors)

## 6.3    Triumphant Populism

There is a type of injustice which rejects all those who do not hold what is considered to be the majority opinion, or rather the way the majority thought two decades ago. The term "populist" is already a rejection in itself, with its negative connotations. In fact, everything we are describing, that feeling of abandonment rather than exclusion, of a loss of bearings, is an obvious fact for many. But our fear is that this de facto abandonment of politics leaves the field clear for those we have called the "technological prophets", leading our society to consider the polarisation of incomes and knowledge as justified in the name of a shining future of innovation.

That is why it is so important to understand what we are dealing with when we talk about populism. This term, which has so many connotations, invites the condemnation and loathing of moralists, who will often recoil from it and throw their hands up in horror. This does not seem to be the way the subject should be approached: it runs the risk of missing the crux of the matter. For this reason, we cannot reduce populism to demagoguery, that political style which the great Ancient Greek authors identified and criticised so well.

The difficulty in defining the people, other than as an abstract entity, applies to the term "populism" as well. There could be as many concepts of the people as there are versions of populism. Thus it could mean, in the current European context, an ethnocentrism which is assumed to be in response to people's fears of immigration. It could also arise from the theory of the authoritarian personality developed by Theodor Adorno, which is characterised by antidemocratic, fascistic behaviour and intolerance of minorities, linked to traditional, closed structures in society. One could also link it to modernisation or to the end of the industrial era and its neglected remnants, like the Rust Belt in the United States. Finally, it is also used to discredit movements as diverse as Syriza in Greece, Cinque Stelle (The Five Star Movement) in Italy and PIS (Prawo I Sprawiedliwosc: Law and Justice Party) in Poland, born out of the frustration of the population, from its feeling of loss of status and from its anxiety about the future. And yet, we know there are also examples of ultra-conservative voting by populations who are living very well.

By populism, then, we should understand a conservative upheaval, which is not afraid to question democratic norms. Populism is therefore an internal issue of democracy and not a type of movement which is foreign to it. It is an issue which must be dealt with seriously; we must avoid stopping at a simple condemnation. It is the response, albeit a simplistic one, to a real distress, a real "social disarray", with, to paraphrase Pierre Rosanvallon, a feeling of powerlessness, an absence of choice and a world which is hard to understand. What Rosanvallon also teaches us is that populism is only a simplification of things: it could not replace a political programme. It is a political simplification to define the people by pitting them against elites; an institutional simplification to avoid all intermediaries and mediation and it is a simplification of social relations to reduce them to an identity defined as anti-immigrant, anti-Islam, etc. However, there is a new element: a kind of return of the least educated, which the study by the Pew Research Centre demonstrates very well. The new rich are among them, taking a stand against a globalised, cosmopolitan, racially diverse, establishment offering equal opportunities for women, along the lines of New York City or the American West Coast. A Harvard Business School paper (Di Tella and Rotemberg 2016) posits the hypothesis that from the least educated people's point of view, it would be better to elect an incompetent who could not be accused of betrayal if incomes fall.

We are seeing it today. Populism seems to be thriving on the profound crisis in democratic institutions, the increasingly significant paralysis, according to Étienne Balibar (2017) of European parliamentary systems and the ungovernability affecting one country after another, whether it be the UK, Italy, Spain or France. Balibar says it is the result of the collapse of the European project. For the United States, it will be the decline of its empire, which could breach the very fabric of the constitution. In short, populism is the expression of, but not the remedy for, the crisis of legitimacy and representation in our systems.

Balibar is far from being the only one to offer such a diagnosis of populism: that attempt at de-democratisation, which not only affects the west, but also India, South Africa and Latin America. For the essayist and historian Timothy Garton Ash, "we face the globalisation of anti-globalisation, a popular front of populists, an Internationale of nationalists" (Garton Ash 2016). The United States, the oldest liberal democracy which respects the balance of powers (the sacrosanct "checks and balances") has just moved on from it.

To illustrate this, the last word goes to Fareed Zakaria, who introduced the idea of "illiberal democracy", an apparent contradiction in terms, two decades ago in *Foreign Affairs* (Zakaria 1997). An illiberal democracy is a democracy, but one which does not respect the state, the law, minorities, freedom of the press, etc. In 2003, he wrote that Americans already showed the greatest respect for three bodies which he judged to be "undemocratic": the Supreme Court, the Federal Reserve and the Army. Today, he views the rise of illiberalism in the United States as a phenomenon which is shared by everyone: Democrats and Republicans. But populism has new roots which explain its rejection of politics. And there are all kinds of prophets who plan to create a universe which is finally purged of politics: of its corruption, of its inability to think of the future and of its powerlessness to act. The technology prophets are the obvious example of this and believe it is legitimate to replace a failed politics. And the movement will not stop at that. This movement will not triumph easily on a political level, but it will undoubtedly be one of the major components of the future structure of our society. Essentially, it is a question of getting past this fluid phase of discredited democracy. Democratic institutions are no longer ultimate points of reference: they are the very opposite. Politicians are no longer the representatives of a social contract that John Locke, Thomas Hobbes or Jean-Jacques Rousseau would have wanted. In a similar way to the Marxists giving the working class the role of Messiah, our current prophets are changing God: from now on it is their technology. All of this will take time: the time for major innovations to happen. But the narrative has already begun. And that is what will shape public debate from now on.

## Notes

1. Mill JS (1848) *Principles of Political Economy*, vol. 1 (2016 ed) CreateSpace Independent Publishing Platform, California.
2. Oxfam (2016) An Economy for the 1%. Available via https://policy-practice.oxfam.org.uk/publications/an-economy-for-the-1-how-privilege-and-power-in-the-economy-drive-extreme-inequ-592643.
3. Pre-revolutionary France.
4. Approach taken by the Pew Research think tank.

# References

Balibar, E. (2017). *'Populism' and 'Counter-Populism' in the Atlantic Mirror*. Open Democracy. Available via https://www.opendemocracy.net/en/can-europe-make-it/populism-and-counter-populism-in-atlantic-mirror/.

Cowen, T. (2013). *Average Is Over, Powering America Beyond the Age of the Great Stagnation*. Boston: Dutton.

Di Tella, R., & Rotemberg, J. J. (2016). *Populism and the Return of the "Paranoid Style": Some Evidence and a Simple Model of Demand for Incompetence as Insurance Against Elite Betrayal* (Working Paper 17-056). Harvard Business School.

Freeland, C. (2013). *The Rise of the New Global Super-Rich*. Ted Conference. Available via https://www.ted.com/talks/chrystia_freeland_the_rise_of_the_new_global_super_rich.

Garton Ash, T. (2016). Populists Are Out to Divide Us. They Must Be Stopped. *The Guardian*. Available via https://www.theguardian.com/commentisfree/2016/nov/11/populists-us.

Lipietz, A. (1998). *La Société en sablier – Le partage du travail contre la déchirure sociale* [Society in an Hourglass—The Sharing of Work Versus Tearing the Social Fabric]. Paris: La Découverte.

L'Observatoire des inégalités. (2018, March). *Riches, pauvres et classes moyennes: comment se situer?* [Rich, Poor and Middle Class—Where Do We Fit In?]. Available via https://www.inegalites.fr/Riches-pauvres-et-classes-moyennes-comment-se-situer.

Marivaux. (1731–1742). *La Vie de Marianne* [The Life of Marianne]. Paris: Le Livre de poche, 2007.

McKinsey. (2016, July). *Poorer Than Their Parents? Flat or Falling Incomes in Advanced Economies*. Available via https://www.mckinsey.com/featured-insights/employment-and-growth/poorer-than-their-parents-a-new-perspective-on-income-inequality.

OECD Focus. (2014). *Top Incomes and Taxation in OECD Countries: Was the Crisis a Game Changer?* Available via http://www.oecd.org/social/OECD2014-FocusOnTopIncomes.pdf.

Oxfam. (2017). *Just Eight Men Own Same Wealth as Half the World*. Press Release. Available via https://www.oxfam.org/en/pressroom/pressreleases/2017-01-16/just-8-men-own-same-wealth-half-world.

Pisani-Ferry, J. (2017). *Le renoncement au progrès*. Project Syndicate. English version available via https://www.project-syndicate.org/commentary/

populism-and-abandonment-of-progress-by-jean-pisani-ferry-2017-01?barrier=accesspaylog.

Stiglitz, J. E. (2015). *When Inequality Kills*. Project Syndicate. Available via https://www.project-syndicate.org/commentary/lower-life-expectancy-white-americans-by-joseph-e-stiglitz-2015-12?barrier=accesspaylog.

Stiglitz, J. E. (2016, March). *The New Generation Gap*. Project Syndicate. Available via https://www.project-syndicate.org/commentary/new-generation-gap-social-injustice-by-joseph-e–stiglitz-2016-03?barrier=accesspaylog.

Tawney, R. H. (1964). *Equality (1931)*. London: Allen and Unwin.

The World Bank. (2016). *Poverty and Shared Prosperity 2016—Taking on Inequality*. Available via http://www.worldbank.org/en/publication/poverty-and-shared-prosperity-2016.

Zakaria, F. (1997, November–December). The Rise of Illiberal Democracy. *Foreign Affairs, 76*(6), 22–43.

# 7

# Who Governs: Politicians, or Technology Prophets?

The beginning of the twenty-first century strangely resembles the end of the nineteenth century, when large trusts, in particular the American oil companies, effectively governed their own country and, to some extent, the world. It also recalls the more recent episode of financial madness which led to Goldman Sachs having a decisive influence over almost all world governments. The real difference between today and the past is simply that the power of the large technology companies is not yet established in practice, although it is in firmly so in people's minds. States are growing weaker; they are struggling to understand the current constraints and a future shaped by ageing and technology, which makes it impossible for them to define the future. This has left an empty space to fill! The large technological companies, whether in digital, genetic, energy or space transport technology, quickly understood this public vacuum, and defined the future—their future, and now ours too—by designing society as they see it: above all, one which would allow them to maintain their power. The best example of all of this is surely Elon Musk's public launch of a hare-brained project: the conquest of Mars. It was strikingly reminiscent of the Space Race between the USSR and the United States, when both countries' leaders took turns to describe

© The Author(s) 2019
J.-H. Lorenzi and M. Berrebi, *Progress or Freedom*,
https://doi.org/10.1007/978-3-030-19594-6_7

the conquest of space in terms of their political objective to build a new frontier.

For the first time, public discourse is more important than reality. As a result, when we wonder about the future, we all feel that radically new technological solutions are within reach. This is obviously entirely in the hands of about a hundred large companies, from all nationalities and sectors. There has never been such a strong feeling that society is dominated by the Monopoly of a few: at least not for one and a half centuries. And these same companies must now fortify their limitless ambition and a power which is still largely in the realm of words. That is the reason why monitoring individuals and states is important to them, because this surveillance actually allows them to assert their power at all levels, and, more strangely, to justify their approach. This chapter will deal with all of this, with a very specific illustration of the development of power relationships between states and these large technological firms. One of the talents of these firms is to avoid the principle of public obligation: taxes. This means avoiding financing all other activities taken care of by the state. This situation will not last: but it remains a unique moment in the relations between players in the economic and political arenas.

## 7.1    The Future of Human Society in the Big Tech Firms' Hands

There are actually a very large number of these firms, even if we sometimes only highlight today's most iconic examples: the digital companies. They are designing what will, in the coming years, be an industrial revolution. It comprises genetic engineering, astrophysics, new forms of energy and changes in patterns of consumption and therefore in objects produced, and also the production processes for these new products. It encompasses all scientific fields and the whole value-added chain. But as we have seen, we are only at the early stages. Nevertheless, these firms affect every country in the world; they will be the astonishing driving force of the coming decades, supported by an acceleration in scientific progress and above all by visible, yet limited, technological upheavals. It is only a question of time, of course; but without disrupting the markets, these hundreds of businesses are designing the world according to their vision right now, and firmly believe that they are seeing it being built in front of our eyes. They are called: Open Bionics, Suntech, Azuri Technologies, Monte Alto Renovable, Carmat, Suzlon Energy, Cheetah Robotics, KrioRus, among others. These names are mostly unknown and yet these companies are the agents of change in all fields: from medicine to transforming energy, by way of astrophysics. They are based everywhere: in the United States, but also in France, the UK, Germany, China and even Chile.

In some ways, they grab our attention with their often worrying predictions. Just listen to Ray Kurzweil in *The Singularity Is Near* (Kurzweil 2006): "By understanding the information processes underlying life, we are starting to learn to reprogram our biology to achieve the virtual elimination of disease, dramatic expansion of human potential, and radical life extension". It is true that the genome will soon hold no more secrets for mankind and this could actually postpone illness and increase life expectancy. But the most troubling innovation to come must be the possibility of defying death via cryonics: the freezing of corpses with the aim of reviving them hundreds of years later. Far from being a vision of science fiction, companies like the Russian KrioRus, or its American counterpart the Cryonics Institute are already offering cryonics to their clients, hoping that in a few decades from now they will be able to make us immortal.

Physicist and Nobel Prize winner for Chemistry, Horst Stömer adds, in the same work: "Nanotechnology has given us the tools which enable us to manipulate atoms and molecules. Everything starts with that. The possibilities for what we can create appear infinite". It is true that the achievements of artificial intelligence are stunning. An illustration of this is the development of the humanoid robot "Sophia" by Hanson Robotics, which displays completely human characteristics—the capacity to exchange ideas, to learn, to feel empathy—to the extent that her creator, Ben Goertzel, says robots will one day be fully fledged citizens, with independent minds, who are active participants in the world.

This is not the only field where advances are as astonishing as they are troubling. The conquest of space is also taking a new turn. Elon Musk, founder of Space X, has announced he will be able to send two tourists into orbit in the near future, adding that it was "an important step in preparing for their ultimate aim of transporting humans to Mars in 2030". So where are the so-called GAFA companies in all this? The acronym is often wrongly used to refer to all large tech firms, but it actually refers to Google, Amazon, Facebook and Apple. Those companies have stood out over the last twenty years because of their financial weight, but also, above all, because of their capacity to monopolise a certain type power—the power of influence and words—without anyone really noticing.

The technology companies' narrative is firstly and most importantly about explaining to the world that their role is to make people's lives easier, to work for the good of humanity and to make this world better. In an environment which is increasingly sceptical about the political classes and elites, they look like the saviours of this rudderless world, ready and able to meet the challenges that governments have not understood or been able to meet.

There are many examples which illustrate this aim. First of all, most technological companies are convinced that the world's violence can be explained in part by a lack of communication. It is a gulf to be spanned by a bridge between all human beings on the planet: the Internet bridge. That is why each of them is developing its own project to connect the world and to modernise remote regions of Africa, Asia or South America in doing so. Google's Project Loon consists of creating an Internet supplied by a network of balloons in the most distant areas of the globe, designed to connect 4 billion people around the world. Facebook, with its Internet.org project, is firmly convinced that "The more we connect, the better it gets".[1] That is why Mark Zuckerberg launched an appeal to computer developers worldwide and has already conceived a way to connect the whole world using a system of drones run on solar energy. As for Elon Musk, his project is called "OneWeb". His idea is to deploy a few thousand satellites, 4425 to be exact, in order to make broadband connection widely available. He plans to supply the United States first of all, then to cover all the regions of the world in a second phase.

The large technology companies are also engaging with the struggle against climate change. The Ivanpah Solar Power Plant, located at the boundary between Nevada and California in the Mojave Desert, is a perfect illustration of this. It is a solar farm capable of supplying 392 megawatts of solar energy, or enough energy to supply around 140,000 households, without emitting carbon dioxide or radioactive waste! Moreover, while Google consumes almost as much energy as a city the size of San Francisco all on its own, particularly because of the dozens of "data centers" scattered all over the world, the company nevertheless ensures that its electricity consumption is "virtually clean". What this means is that it buys the equivalent of its electricity consumption from producers of wind and solar energy, namely 2.6 gigawatts, which makes it the biggest buyer of clean energy in the world. Moreover, it boasts on its blog about the size of this figure, because it is more significant that most electricity suppliers and also because it is twice the size of the famous 1.21 gigawatts Marty McFly needed to get home in the film *Back to the Future*. For its part, Apple announced in 2015 that it would build two "data centers", one in Ireland and the second in Denmark, which would be supplied entirely by renewable energy. It is the same story for Facebook and Amazon, who have quite similar objectives, and who are aiming for a clean energy supply. When it comes to the transition to new energy sources, other companies have made green growth their credo. They are following in the footsteps of the British firm Azuri Technologies, which markets solar kits which give rural areas in Africa access to energy.

The large technological companies also appear to be the solution to making health care accessible to all. According to a recent study (Deloitte 2016), 59% of French people, of which 68% were aged between 18 and 24, trusted the large technological companies on the issue of prevention. With Verily,[2] one of the Google's so-called moonshots, or utopian projects, the company wants to develop initiatives to guarantee better health and save lives. For example, it is planning a contact lens which would enable diabetic patients' glucose levels to be checked.[3] It is also working on a surgical robots platform[4] which plans to develop a pill made up of nanoparticles which would circulate through the bloodstream, detect potentially harmful substances and stick to them.[5] It has also come up with connected cutlery,[6] to make mealtimes easier for people with Parkinson's disease by minimising their trembling.

As far as the issue of road traffic goes, the various self-driving car projects appear to be the solution. The promise made this time is not just that road traffic will be optimised, thanks to cars which never cause accidents, but also that it predicts a major change in consumer behaviour. Drivers, who aren't really drivers any more, could now make use of their journey time to take part in various activities or hobbies and thus get the most out of their travel time, rather than devote the time to driving. Bad news for the many people who liked driving: the big technology companies have got their wellbeing in mind.

The large technology firms are even studying possible solutions to the problem of terrorism. One group among them has announced an unprecedented partnership with the aim of fighting against cyber jihadism and promoting a "counter-narrative" in the face of Islamist propaganda. This was how a "Safety Check" button was developed, which enables members who are detected to be in danger zones to reassure their loved ones and to confirm they are safe and sound in case of emergencies, whether that be terrorist acts or natural catastrophes.

In a world which is: troubled by insurmountable inequality, traumatised by the fast track period of de-industrialisation at the end of the 1990s and the consequences of globalism of the 2000s, worried about the extent of climate change and natural disasters, destabilised by the terrorist threat and more and more tempted by insular identity politics, one could believe that the big technology companies are in a position to find suitable solutions. These companies are gradually supplanting governments by offering solutions and answers to the problems which states have certainly not managed to resolve.

We could almost be witnessing the move from a world ruled by civil servants and technocrats towards one monitored and designed by highly gifted engineers. And in this newly defined world, Silicon Valley represents the breeding ground for intelligence and innovation. Silicon Valley crystallises that image of the Earthly Paradise: Tech Eden. It is a land where the sun shines almost all year round, whose inhabitants are at work trying to solve human misery and to do good. And the model extolled by this Californian Eldorado exerts a real force of attraction. Whether it is young graduates from the world's most prestigious universities, or investors and other capital-risk funds, Silicon Valley fascinates and generates enthusiasm everywhere. The values of tolerance and creativity which this Eden celebrates, the desire to act collaboratively, intelligently, for the good of mankind; the incredible success, the assumed friendliness; the optimism about seizing the future; its cool attitude which sets creative minds free… there are so many examples available to try to explain this global fascination with the Californian model! But, of course California has spawned some imitators. Gradually, new technology hubs have emerged all over the world. France promotes French Tech, Singapore defends its entrepreneur and innovation support "ecosystem" and Israel highlights its Silicon Wadi, which in the space of ten years has enabled more than 4000 start-ups. These include Waze, which was bought by Google, and Viber, bought by Rakuten. In Dubai, they talk about the "Silicon Oasis". In South America, people often talk about Santiago de Chile or Argentina's "Silicon Barrio". In South Africa, there is "Silicon Cape Town"; in India, the City of Bangalore. China has the Zhongguancun district and Taiwan has Hsinchu. In South Korea, there is Incheon; in Australia, there is "Silicon Beach"—the list is endless!

Eric Sadin says (Sadin 2016) the term colonisation can be used to describe the impact the spirit of Silicon Valley has on the minds of the rest of the world. He even talks about siliconisation, a sort of colonisation which is however different because it is not experienced as violence to be endured, but rather as "an aspiration, ardently wished-for by those who intend to submit to it".

Omnipresent in universities, at major international summits and in the press, the significant impact of the large technological companies is measured by their impact on public discourse. And it must be said that these businesses effortlessly took over the calming narratives that surrounded the birth and development of the Internet. Like missionaries for American-style democracy, they boasted of a technology which was accessible to all, which made all human beings equal to each other, a smooth-running world, rid of wars at last: a global village, to use McLuhan's expression. It was as though technology assumed a moral dimension and bestowed on everyone not just access to happiness, but what is more, the right to happiness. This "civilising" mission which knows no borders, which overcomes constraints which could be presented by language, religion or history; which plays at concocting ideas on a planetary scale could raise a smile when we look at the current dark state of the world. But although these heroes of technology (because this goes well beyond digital technology) think of themselves as serving a saintly ideology, of playing with words and meanings, their efforts cannot compare to the virtuosity and intelligence of the ancient sophists. The sophists were criticised heavily by Plato and Aristotle for their "apparent wisdom", but they were none the less very fine rhetoricians. Those heirs of Protagoras turned the speeches they made in the course of discussion and debate into instruments of power and used them to defend their interests.

We have come a long way from the few "universalist" and simplistic messages of the promoters of the large technology companies. And yet, it is a matter of power, of spreading an ideology which is shared by the Silicon Valley communities and now set up as a model to follow by the connected human race. There is within it a kind of circumvention of politics and of its institutions, in favour of the self-establishment of an order defined by technology, whose consequences we are a far from having experienced. It could be argued that the narrative of the large technology companies, as poor as it is, has something imperial, if not imperialist about it. Which reminds us a little of Paul Virilio, a philosopher who pays such close attention to the phenomenon of speed. "If anything is imperial, it is definitely the information super highways […] I recall that the source of the Internet is Arpanet, the network of networks of people who carried out research in the military field. Why hide it? The superhighway and the hidden source of the internet go together too well" (Le Monde 1997).[7]

It also happens that this imperial position is similar in substance to the current American national identity. But we can already see the Chinese equivalent appearing: our technological apostles hold a very similar view to the first colonisers: the Puritans who exiled themselves from England to flee repression, but also to leave the Old Continent, which they considered decadent. Just like John Winthrop, the first governor of the Massachusetts Bay Colony, they share that same faith in being a "chosen people", and the same mission: to save the world. Winthrop seemed to be inhabited by that conviction in 1630, before he had even set foot on the land, when he announced (Winthrop 1630) to his fellow travellers: "we shall be as a city upon a hill. The eyes of all people are upon us, so that if we shall deal falsely with our God in this work we have undertaken, […] we shall be made a story and a byword through the world". This messianic attitude towards the other nations of the world, born out of a puritan Protestantism, seems to still be present in the way the United States talks about the rest of humanity, first during their territorial conquest of the United States and then in their foreign policy. That "Manifest Destiny" is always the justification for their territorial expansion and their mission to fight and punish the forces of Evil. The United States owes the idea of "Manifest Destiny", which fast became its most deeply held credo, to John Louis O'Sullivan, the director of the *Democratic Review*. In 1845, the date when Texas was annexed, he wrote: "America was chosen for this holy mission to the nations of the world, who are deprived of the life-giving light of the truth […]".[8] The public message of the large technology companies is not unlike this "Manifest Destiny", promising the best of worlds to humankind.

Is the power of the large technology companies really capable of rivalling the power of states? Nothing is more certain. Let's remember that unlike traditional states, large technology companies do not have borders. Their power extends everywhere around the world and they manage to gradually increase their users' dependence on the innumerable services they offer all the time. Their financial weight is formidable, and these groups are complete masters of tax avoidance techniques so that they can circumvent taxes. They are not yet at the stage of writing laws, but as Harvard law professor Lawrence Lessig observes so well: "Code Is Law" (Lessig 2000). Nowadays we are living in a world where the regulator is the code: no longer the code of law, but actual computer code.

## 7.2   A Society Controlled by the Big Tech Firms

It is difficult to define what can be considered the control of individuals by companies. It is obviously about limiting opportunities for criticism, alienation and individuals acting within the framework of a predetermined technological environment. That particular point alone leads us to mention the large technology firms' pioneering role, in which they assign themselves the mission of making life easier for people. And behind that grand ambition, there is data: collecting, processing and storing data reveals itself to be the key resource for supplying economic development. And unlike fossil fuels, data is an inexhaustible resource. The Internet of Things makes each fact and each gesture equivalent to yet more information to feed the predictive models of the algorithms.

How do they go about convincing populations of every socio-professional class to continue to hand over their personal information? How do they encourage men and women to divulge so much private data, when they all know the data is liable to be used for commercial ends?

In order to pull off this masterstroke, businesses have resorted to every conceivable incentive. Tristan Harris was a Design Ethicist. He describes how he operated in his network business. His role was to come up with sources of distraction for the general public. In order to do that, he drew on his experience as a magician. A good magician starts off his performance by analysing the weaknesses and vulnerabilities of his audience, in order to be able to distract its attention and therefore manipulate it without it knowing. Most of today's websites and applications use these techniques in order to keep users on their services for the maximum amount of time.

Let's examine Harris' description, which illustrates his skilfully structured manipulation methods. First of all, in order to control and manipulate his audience, a magician must give the impression that he is offering them the chance to choose. The audience must retain free choice. Except that this is nothing but an illusion; because even if it is for the audience to choose which card to hold, in the end the magician organises the choices on offer, making sure that he always wins. Almost the same technique applies when someone uses their smartphone to search for a place, for example, a restaurant. The application presents the user with a certain number of restaurants which are geolocated to be nearby and the user just has to choose between the ones on offer, in a similar way to having to pick just one card offered by a magician. And he or she ends up choosing one of the restaurants which appears on the phone, without even taking the trouble to look around and check if this choice corresponds well with what he or she wanted, taking into account the other restaurants which are physically present. More generally, menus and drop-down lists amount to an illusion of free choice in the end.

The second principle for encouraging users to stay connected is to make them dependent by developing a craving for gambling and chance. To illustrate this, Harris makes a comparison between smartphone applications and slot machines. Slot machine addiction depends on two types of uncertainty: firstly, the fact that when one plays, one never knows what will happen; and secondly, what the potential gain will be if one happens on the right set of symbols. The principle is almost the same when users continually refresh their notifications or when they swipe their screen with their fingertips to scroll through users' profiles, whether for professional applications or recreational websites. They never know what or whom they will come across. And that is what makes the experience so addictive. Added to that are the fear of missing an important event and the fear of being out of touch. This fear encourages people to remain subscribed to newsletters they no longer read, to follow certain news feeds they are no longer interested in and to remain friends on social networks with people they no longer have regular contact with. "The demand on the user to put their social world on to online networks is now a response to the fear of being cut off, left out of the current world of communication and society" (Casilli 2010). Whether you call them digital natives or Generation Y (Dagnaud 2011), many surveys and testimonies describe the strength of these digital relationships which push this new generation to be permanently in the process of writing, sending or receiving messages; enclosed in a continuous stream of conversation, from morning till night, only to start again when they wake the next day. And alongside that fear of being out of touch, there is the incentive to be connected fuelled by the phenomenon of permanent personal self-expression. Nowadays we have to be visible by everyone, everywhere, all the time. We maintain Facebook pages, post photos on Instagram accounts, and look at others on Snapchat. We reveal ourselves and confess secrets, relinquishing our anonymity. For Geert Lovink, media theoretician, "people think that their freedom demands of them that they 'speak the truth', that they confess to someone - a priest, a psychoanalyst or a blog - and the fact of speaking that truth will set them free" (Lechner 2008).[9] Diana Tamir and Jason Mitchell from Harvard University (Tamir and Mitchell 2012) explain that this enables our brain to release dopamine, the desire molecule. Now, in a normal conversation, researchers estimate the time spent talking about one's personal experiences is between 30 and 40%, whereas on social networks like Facebook, it represents 80% of our activity.

Another incentive for people to stay online is that content is engineered to be endless and unlimited. When you click on a video on YouTube or Netflix, a new video starts up automatically when the first one finishes, pushing the user to spend longer online. When you scroll through your newsfeed on Facebook or LinkedIn, the page continues to load and supply new information to the news feed. It is essentially the same principle as the study by Brian Wansink where he showed that it is possible to make people consume 73% more calories if you serve their food in bottomless bowls (Wansink et al. 2005). Consumption keeps up with whatever quantity there is to eat.

The impression of urgency that smartphone applications provoke in their users, the notifications which appear on screens, the instant messages which show correspondents that the message sent a few seconds earlier has actually been read, the latest news alerts, which appear automatically… all these elements burst into our daily lives and reinforce our feeling of dependence on our devices. Even better, some sites gather together more and more services, which previously had nothing in common, with the sole aim of obliging us to pass through their network. We will be forced to go via a newsfeed on a site which we normally use to check on events or find a friend. And when a user passes through the newsfeed, he or she is exposed to product placement, selectively offered, based on the users' profile, location, the time of day, etc.

There are in fact many techniques for hooking users on to a digital drip. And in making themselves indispensable, the tech firms have pulled off the impossible feat of simultaneously asserting their control over users and over many other areas of activity. But what is most striking is the monopolistic nature of the provision of these services. The turnover of four large digital companies exceeds the GDP of individual countries such as Denmark, Norway or South Africa, while their salaried workforce is incomparably smaller. By gradually taking control of society, they have therefore ended up rivalling the power of the states they initially charmed and circumvented. It has always been the wish of some government agencies, such as NASA, to support the emergence of a digital strategy for the United States. Relations between the large technology companies and the American government are based, as one would think, on reciprocal interests. On the one hand, the United States wants the development of businesses like the large technology companies to contribute to its international standing, helping them to retain their leading place at the frontiers of technology. On the other, the large technology companies need the goodwill of public authorities to be able to develop their new activities.

The strongest sign of the new-found importance the large technology companies have acquired may be that a country like Denmark wants to send an ambassador to these companies. The proposal from the foreign minister shows that Denmark is ready to consider that the large technology companies should enjoy the same status as a nation state. They are no longer just economic entities; they now embody real political power. In the end, it could be said that the large technology companies' narrative has ultimately convinced the governments themselves of the companies' importance. The ultimate absurdity, for the moment, is that the large technology companies show their new power relationship with states via their will and their ability to avoid state taxation.

## 7.3   The Dream of Avoiding State Taxation

This may be the area where the new reality of this power relationship between the large technology companies and states is most significant in understanding the companies' final objective. What could be more symbolic, in the context of the type of global market predicted for the coming decades, than the power relationship between states and the large technology companies, which is transforming itself to the companies' advantage, so that they are exempt from taking part in contributing to common-pool resources?

What appears today to be the simple implementation of tax expertise, limited to a moment of historic weakness by the states, is in fact very important for considering the future. We are not setting out to stigmatise the large technology companies, who are without doubt an exceptional creative force, but rather to analyse a power relationship which will pose a major problem in the coming years.

The issue is in reality the difficulty that states, which are isolated, have in constraining companies, who are obviously extraterritorial. They are like science: universal; and as such, they refuse local taxes that they judge to be outmoded and unconscionable in a global world. In fact, everything goes back to the problem of double taxation.

It was with the development of international trading and the boom in people's mobility that the problem of double taxation appeared, at the very beginning of the twentieth century. From the 1920s onwards the League of Nations, the forerunner of the UN, tackles the issue. An initial committee of tax experts, made up of Professors Bruins, Einaudi, Seligman and Sir Joshua Stamp, presents a report to the Genoa conference in 1922 which proposes several approaches to resolving the issue of double taxation and puts together the initial foundations for an international tax system. For countries who are members of the OECD, it is important to retain two fundamental principles of international taxation. Firstly, the notion of permanent establishment, which means a company, is only taxable in the country where it "exercises its activity". Secondly, the principle of arm's length pricing, which means considering subsidiaries of a group as being independent entities. Thus, the price charged between dependent business enterprises should be the same as that which would be applied between two independent businesses.

This made sense until new sectors of activity emerged, particularly the digital sector, and the structure of businesses changed, with some of them managing to circumvent the international tax rules in a very skilled way. We are not talking about tax evasion here, which is an illegal activity, but rather tax avoidance, which consists of managing the laws to minimise tax rates. The main method used to minimise companies' tax rates is via the transfer value. This is not new. It concerns the price charged when a product is sold within a company or between two subsidiaries. According to the OECD, the price applied is supposed to be an "arm's length" price and therefore in line with the market price. Except that the notion of "arm's length" can sometimes reveal itself to be quite fluid, particularly when no equivalent products exist on the market, or when it concerns a patent which is on loan but has not been handed over.

Let's take the following example. Let's suppose a parent company sells the usage rights for its patents for a very low sum to a subsidiary located in a tax haven. Then suppose that the subsidiary located in the tax haven then trades the rights to exploit the patents to a regional subsidiary, this time for the entire profit generated. The regional subsidiary can then exploit the rights, but it can also surrender them to other local entities, against almost the entire profits generated. With this plan, neither the local subsidiaries nor the regional subsidiaries make profits because most of the profits are transferred to a tax haven in exchange for the right to exploit patents. And in this way, the subsidiary located in the tax haven (which therefore deducts almost no taxes) makes very significant profits. This whole scheme existed before the beginning of the twentieth century. There was a famous example of the brothers William and Edmund Vestey, leaders of a meat empire, who took care to concentrate their profits in a country with low taxation at the time: Argentina.

A newer phenomenon is linked to the fact that certain countries sometimes decide to tax these transfer prices when intra-group transfers are made with foreign entities. This gives rise to another tax avoidance technique, nicknamed either the "Double Irish" or the "Dutch Sandwich" method. Ireland is different from the other members of the European Union in that it does not tax Irish companies located abroad or payments transferred out of the European Union within Irish companies. This tax avoidance strategy involves three different countries: Ireland, the Netherlands and a tax haven like Bermuda, the Caiman Islands or Gibraltar, where there is almost no tax. The tax avoidance system is as follows: the American parent company sells the rights to exploit intangible assets outside of the United States to an subsidiary governed by Irish law, whose functions are performed by a permanent establishment located in a tax haven. In this way, all the profits made outside of the United States are declared by that subsidiary and thus not taxed by the American tax administration. At the same time, the Irish company controls another subsidiary, which is also based in Ireland—hence the term "double Irish"—whose role is to make sales outside of the United States and to record the related turnover. This second subsidiary then has its turnover cancelled by paying an intellectual property licence fee to the parent company's permanent establishment, situated in the tax haven. In order to optimise the performance of this system, the licence fee can pass through the Netherlands—hence the term "Dutch Sandwich". Passing through a notional company with no employees, a kind of postbox based in the Netherlands, allows the company to enjoy favourable terms concluded between Ireland and the Netherlands, and in particular to benefit from the fact that there is no deduction at source, even if the state where the counterpart is based is a tax haven.

In October 2010, Bloomberg revealed how Google used that technique to pay just 2.4% tax on its profits outside of the United States. According to a study by Citizens for Tax Justice and the US Public Interest Research Group, more than 2100 billion dollars of profits accumulated abroad by the 500 biggest American companies were kept abroad by them at the end of 2014. The most iconic of those companies would be Apple, which held on to 181 billion dollars offshore. The large technology companies primarily devote that money to making foreign acquisitions, and therefore to ensuring their dominance of the market is even more entrenched against any possible future competition which could emerge from a young start-up.

The issue of tax exile has become a particularly burning one in that the loss of income for governments is made up by everyone else: the taxpayers. The problem for states nowadays has moved on from avoiding double taxation to actually avoiding double exemptions for companies. Some countries have not hesitated to try unilateral approaches, along the lines of the UK or Australia, who have already instituted a tax on "diverted profits". In order to put an end to these practices which consist of declaring artificial profits in countries with low taxation when the actual turnover is made elsewhere, the British government decided to tax at 25% all "diverted profits" instead of the 20% charged, and in doing so hoped to discourage bad practice. It remains to be seen how to identify these "diverted profits". The real innovation here lies in the power of the British government to force the payment of this tax, which many have nicknamed the "Google tax".

Tax has always had an unbreakable link with citizenship. This link relies on an idea which is basically quite simple: that states have to create an environment and protection which are beneficial to the population and that the spending associated with that is financed by taxes. And wherever we are located in the world, that perspective remains the same. Living together is therefore at the heart of that tax. Of course, we are at a period in history where there is a widespread feeling that we should reduce people's obligations, very simply because we are in a phase of individualisation. And that is what is at stake. Faced with the evocative and transformative power of science, individuals find themselves effectively confronted with these technological firms as powerful advocates, claiming to provide as much well-being as states. Is this true or not? The question deserves at least to be put and power politics should at least be defended. Europe, finding itself so disrupted, has attempted to assert that position and even to base its will to exist on the will to fight against tax avoidance. This is why the European Commission proposed the relaunch of the Common Consolidated Corporate Tax Base in 2016.[10] This project's ambition is to impose the same rules for calculating companies' taxable income, the same mechanisms for consolidating profits and losses, for all of the European Union member countries. However, such a project requires the unanimous agreement of all the EU members in order to be adopted, which of course will not happen. What we should take away from this is that the technical tax base problem is a truly political response to the attempts by the large technology companies to avoid tax. But fortunately, Europe is not alone.

The OECD has drawn up a plan called Base Erosion and Profit Shifting (BEPS). The idea for such a project emerged around 2012. The OECD then obtained a mandate from the G20 in 2013, at the time of the Starbucks affair in the UK, to get this large-scale project, which goes well beyond the thirty-five OECD members, moving. Until now, the OECD has settled for providing principles for interpreting some tax rules. From now on, its challenge really consists of revising existing tax instruments and preventing aggressive tax avoidance. The "inclusive framework" brings together more than eighty countries in 2017, all committed to applying the BEPS measures and to monitoring their implementation. One of the most significant measures, perhaps, consists of obliging companies which have a turnover of more than 750 million euros to declare their activity in each country and to carry out country-by-country reporting (CBCR) to national tax offices, which could in turn send the reports to other administrations. That is not enough for Paul Krugman and Joseph Stiglitz. They argue that we must consider an overhaul of the business taxation system. Stiglitz and Krugman say we must implement a "minimum corporate tax". That would mean that Apple, for example, which earns 35% of its profits in the United States, would, in this scenario, be taxed on that sum and also in the same way in every other country. For Stiglitz, this method of taxation is "the only means of stopping this race towards legal and organised tax evasion".[11] States will not be able to recover their power to act until they manage to reign in this fiscal dumping, and, in the first instance, the large technology companies.

The winner of this challenge will not only find extra room for manoeuvre, they will also recover some degree of freedom. But the victory itself will also be highly symbolic and will determine, in some way, the future of generations to come.

## Notes

1. Internet.org. Available via https://info.internet.org/en/mission/.
2. Previously Google Life Science.
3. In partnership with Alcon, the ophthalmic subsidiary of the Novartis group.
4. In partnership with Ethicon, a subsidiary of the Johnson & Johnson pharmaceutical group.

5. The Nanoparticle Platform.
6. Liftware.
7. Interview with Paul Virilio, Le Monde, January 7, 1997.
8. O'Sullivan J (1845) *The United States Magazine and Democratic Review*, VI, pp. 426-438, In: Johannsen RW (1997) *Manifest Destiny and Empire*. Ed Haynes SW & Morris C. Arlington, Texas.
9. Quoted by Lechner M (2008) L'anonymat n'est plus qu'une notion nostalgique (Anonymity Is Just a Nostalgic Notion Now), in *Liberation* January 12. Available via https://www.liberation.fr/week-end/2008/01/12/l-anonymat-n-est-plus-qu-une-notion-nostalgique_62540.
10. CCCTB: Common Consolidated Corporate Tax Base. See https://ec.europa.eu/taxation_customs/business/company-tax/common-consolidated-corporate-tax-base-ccctb_en.
11. Proposal by Stiglitz at the 10th Trento International Economics Festival: Des Prix Nobel pour contrer les pratiques d'optimisation fiscale (Nobel Laureates Against the Practices of Tax Avoidance). Sandro Faes, RTBF, June 3, 2015. Available via https://www.rtbf.be/info/economie/detail_des-prix-nobel-pour-contrer-les-pratiques-d-optimisation-fiscale?id=8997589.

# References

Casilli, A. (2010). *"Petites boites" et individualisme en reseaux: les usages socialisant du Web en débat* ["Little Boxes" and Individualism in Online Networks: A Discussion of the Socialising Uses of the Web] (pp. 54–59). Réalités industrielles.

Dagnaud, M. (2011). *Génération Y: Les Jeunes et les réseaux sociaux, de la dérision à la subversion* [Generation Y: Young People and Social Networks, from Derision to Subversion]. Presses de Sciences-Po, coll. "Nouveaux débats".

Deloitte. (2016). *Les Français et la santé*. Etude Santé. Available via https://www2.deloitte.com/fr/fr/pages/sante-et-sciences-de-la-vie/articles/les-francais-et-la-sante-etude-2016.html#.

Kurzweil, R. (2006). *The Singularity Is Near*. London: Penguin.

Lessig, L. (2000). Code Is Law: On Liberty in Cyberspace. *Harvard Magazine*. Available via https://harvardmagazine.com/2000/01/code-is-law-html.

Sadin, E. (2016). *La Siliconisation du monde, L'irrésistible expansion dulibéralisme numérique*. Paris: Éditions L'échappée.

Tamir, D. I., & Mitchell, J. P. (2012, May). Disclosing Information About the Self Is Intrinsically Rewarding. *Proceedings of the National Academy of Sciences, 109*(21), 8038–8043.

Wansink, B., Painter, J. E., & North, J. (2005, January). Bottomless Bowls: Why Visual Cues of Portion Size May Influence Intake. *Obesity Research, 12*(1), 93–100.

Winthrop, J. (1630). A Model of Christian Charity. In *Massachusetts Historical Society Collections*, 1838, 3rd series (pp. 333–348).

# 8

# Two Possible Paths: The Great Parting of Ways

This is a difficult exercise, describing two possible alternative futures, but it had to be done. We are convinced (it is the purpose of this book) that the twenty-first century will be one of confrontation between public authorities—states, political unions, regions, even cities—and the major private stakeholders of the ongoing technological revolution, of which the new digital world is only a very first step. Who will carry off the victory: the genetic engineering companies or the states? Nobody can predict that today. However, what is certain is that those stakeholders will produce one of two possible societies: two worlds which look almost nothing like each other. This vision of the future has prompted us to create a caricatured version of the choice facing societies who believe and declare themselves to be democratic, but who currently live in an atmosphere of uncertainty, indeed of decline, incapable of taking control of their technological future. Instead of launching into a description of the two possible worlds (a very questionable exercise by its nature, which could exaggerate the dominance of some alternatives or the duration of others), we preferred to engage in an unusual, tricky task: to return to our literary and philosophical heritage and give it a new twist, while remaining true to the original texts.

© The Author(s) 2019
J.-H. Lorenzi and M. Berrebi, *Progress or Freedom*,
https://doi.org/10.1007/978-3-030-19594-6_8

We have seen that when we talk about the power of the major digital or genetic stakeholders, we cannot help being paralysed by fear at the idea of the technological society. We are paralysed by terrifying visions of what it could produce, but also by our inability to conceive of the structure and rules of such a society. We have the same attitude when we imagine a new, worldwide, public organisation which would entail, in practice: new regulations, the definition of strong collective values, and relationships between the responsible political powers that inclined towards co-operation rather than conflict. This undertaking is made even more difficult by the lack of a vision of peaceful, blossoming, collaborative societies on offer in our early twenty-first-century world.

In order to describe those two paths and their possible outcomes, we have therefore enlisted the help of authors whose writing has explored these issues by focussing on the essential features of human history. We have selected four works on the triumph of technological prophecies: George Orwell's classic, *Nineteen Eighty-Four* (Orwell 1949), to illustrate what systematic intrusion in people's private lives could mean; *The Handmaid's Tale* (Atwood 1985), by the Canadian novelist Margaret Atwood, to explore how segregation would work in these prophecies, not just socially, but also between genders; *Brave New World* (Huxley 1932), by Aldous Huxley, to describe the possible results of genetic manipulation and finally, *The Caves of Steel* (Asimov 1954), by Isaac Asimov, to denounce the potential violence of this new technological era. The new order anticipated using these fictional building blocks may be nothing but pure fiction, but we must still pay heed to it and, most importantly, escape it. We have named this hypothetical world "Brave Westworld": "Brave" from Huxley's still-dazzling visionary novel, *Brave New World* and "Westworld" from the television series watched by millions around the world in 2016, with Jonathan Nolan's terrifying description of a futuristic society, where humans and robots wage a merciless war. The second scenario, "Anthropos", envisages a peaceful world: a utopia. It is a world designed for us by Aristotle's Ethics, Locke's Rule of Law, Rousseau's Social Contract and finally, Kant's Perpetual Peace. This is not about returning to the Garden of Eden, but conceiving of a world which respects human beings: where politics has regained all of its ambition and its proper place. There is no prohibition against thinking that an evil course of events can be changed.

Guided by these universal novelists and thinkers, we have constructed two scenarios, by taking liberties and creating pale imitations of the most remarkable of works. Our method may be questionable, but it does however allow us to free ourselves from a dizzying, confused present, to remember that history is one of the most powerful resources for understanding the present and imagining the future and that writing has been and still is the tool *par excellence* for making this world, in all its complexity, understandable.

## 8.1  Brave Westworld

Intrusion; segregation; biological engineering; violence. These words, which resonate throughout the recent and distant history of mankind, reflect the desire for power and control, but also that eternal human dream: to be our own masters and achieve eternal life.

Let's turn to Aldous Huxley and enter his Brave New World.

"Tall and rather thin but upright, the Director advanced into the room. He had a long chin and big, rather prominent teeth, just covered, when he was not talking, by his full, floridly curved lips. Old, young? Thirty? Fifty? Fifty-five? It was hard to say. And anyhow the question didn't arise; in this year of stability, A.F. 632, it didn't occur to you to ask it.

"'I shall begin at the beginning,' said the DHC,[1] and the more zealous students recorded his intention in their note-books: Begin at the beginning. 'These,' he waved his hand, 'are the incubators.' And opening an insulated door he showed them racks upon racks of numbered test-tubes. 'The week's supply of ova. Kept,' he explained, 'at blood heat; whereas the male gametes,' and here he opened another door, 'they have to be kept at thirty-five instead of thirty-seven. Full blood heat sterilizes.' Rams wrapped in thermogene beget no lambs."

Let's continue this scene, not in 632 A.F., but in 2025.

"Still leaning against the incubators, he gave them, while the pencils scurried illegibly across the pages, a brief description of the process known as CRISPR-Cas9. 'This is already an ancient process, but one which changed the world. Clustered Regularly Interspaced Short Palindromic Repeats are families of repeating DNA sequences.' The silence from the pencils was deafening. The director attempted to make himself understood as quickly as he could. 'CRISPR-Cas9 is an enzyme which is capable of detecting a specific part of DNA and destroying it. Do you understand what this means?' The poised pencils waited for the answer. 'Yes indeed, this process of cutting out and propagating DNA enables us to intervene at the embryonic stage to improve or impair the future human being.' One pencil stopped its frenzied scribbling: 'Why impair it?' The director almost choked. 'Must I draw you a picture? This process must be used sparingly, or rather with prudence.' The pencils underlined the word 'prudence', in bold. 'With prudence,' the DHC repeated, 'because all humans, or at least embryos, must only undergo this process according to the rank they are called to occupy in society. Alphas and Betas are the first to have their capacities improved, Deltas and Epsilons are excluded. What I mean by that is that their embryos are Crispered to cut away their predisposition to be equal to the superior classes.' The director knew he had gone too far; but how to make these raw recruits understand the way the world worked?"

We admit this is a strange exercise. But it illustrates quite well that the dream—or nightmare, depending on your point of view—which could be realised via genetic engineering is not new. It is a dream which may tend to dress itself up as reality today, at least according to the prophets of the "enhanced" human, to put it mildly. These are the disciples of transhumanism and post-humanism who, as we said, have chosen to live either in Silicon Valley (the more talkative ones), or in Russia or China (those who prefer shadows and obscurity). This dream, which these experts promote to us under the guise of a humanity which is more "human"—fully realised at last, almost deified—is not good at hiding its corrupt ideology. Who else are they aiming their message at, if not at their small, powerful, super-rich community and perhaps at a few battalions of women and men who are seduced by, or even converts to this religion of the superman? There would be losers, and many of them, under this banishment of humanism as we know and cherish it. Whoever talks about "enhanced man" also refers to all those left behind by these enhancements: which would be most of the human race. We must act urgently. Not to silence them, but to look at them afresh and keep them in their place as sorcerers' apprentices: those who only exist to the extent that they have the ear of the powerful and the power of money. There, we have the first defining feature of Brave Westworld.

Let's go further with this exercise and step into the surveillance society, where intrusion is widespread, described with such virtuosity by that master of science fiction and dystopia, George Orwell. There is a surprising parallel between the world where Big Brother follows every individual's every step, and the one we live in with the new master we chose for ourselves: the smartphone. It may be fun and user-friendly, but it turns out it is a formidable spy. It records our conversations and our web preferences second by second. What if, as we might assume, Big Data came to process and manage that enormous amount of information? Who would benefit?

*He went slowly, resting several times on the way. On each landing, the enormous eye stared at you from a screen. It was one of those infographics which are so contrived that the eyes follow you about when you move. BIG DATA IS WATCHING YOU, the caption beneath it ran.*

*Inside the flat his smartphone's fruity voice was reading out the day's news. Winston did not listen; he had no interest in it. He had no way of silencing it. It was switched on 24 hours out of 24 and was a real spy, capable of informing BIG DATA of any inappropriate gesture in a microsecond. It was impossible to get away from it: you would be subjected to harmful ultrasound if you did. Winston turned a switch and the voice sank somewhat. He moved over to the window: a smallish, frail figure, the meagreness of his body merely emphasised by the blue overalls which were the uniform of BIG DATA employees.*

*Outside, even through the shut window-pane, the world looked cold. Down in the street little eddies of wind were whirling white vapour from air-conditioning and torn plastic into spirals and though the sun was shining and the sky a harsh blue, there seemed to be no colour in anything, except the giant screens that were placed everywhere. The huge eye stared at you from every commanding corner. There was one on the house-front immediately opposite. BIG DATA IS WATCHING YOU, the caption repeated, while the giant eye looked deep into Winston's own. Down at the street level another screen alternated the eye with the words "Brave Westworld". In the far distance, a drone skimmed down between the roofs, hovered for an instant like a bluebottle, then nosedived towards a potential suspect. It was the drone patrol, checking everything was in order. The patrols did not matter, however. Only the Words and Gestures Police mattered.*

This illustrates the relevance but also the ambivalence of Orwell's words. The networked world appears to have put on a friendly face of dialogue, uninterrupted communication and widespread access to knowledge: in short of human advancement. Deep down, the essential absurdity of what our prophets are forecasting is no less than a return to a former world where there would be complete, lasting equality in what could be a sort of technological communism. Through communication, a kind of internalised tyranny, everyone talks to everyone else, handing over their hopes and dreams in an embellishment of access, for everyone, to everything and to everyone. Once again science fiction makes the mask slip and refutes the supposed benefits of technology for mankind. That is the second defining feature of Brave Westworld. We know that it is already the case that fully connected communication is addressed at a future community which is very certainly fragmented: it is divided, fractured and has a permanent hierarchy. There is no faster way to inequality than by taking control of communication: that has been known for a long time. And its complete, total mastery could make future societies face real dictatorship, with its oligarchy and exclusions of every kind. Evil, the "banality of evil" as Arendt denounced it, does not take on the appearance of breaking the law, or disobedience. It is so simple! "Evil" itself will also progress behind a mask. The dictator does not "want" evil. He lives in a bubble and exempts himself from the law, subordinating any respect he owes it to his own self-regard. This exchange, this fraudulent sort of conformity to the law, belongs to the realms of lies, deception, and, the trump card: self-delusion. It turns its own fiction into reality and seduces people, as Arendt describes, with the coherence of its organisation, which does not tolerate accidents. And nothing is more worrying, more dangerous for some people, than unpredictability!

By "communication", however, we must obviously understand the networks which support it, the tools which control it, the technology which, in this highly important field of shared intelligence, could take power today. Margaret Atwood portrays, in riveting style, a world revisited by radical puritanism, obviously outside any present or future reality, where technology can serve the most absolute form of sexism and the domination of women by men:

*I'll come over, she said. She must have been able to tell from my voice that this was what I wanted.*

*She got there after some time… So, she said… Tell me…*

*I tried to tell her what had happened to me. When I'd finished, she said, Tried getting anything on your Compucard today?*

*Yes, I said. I told her about that too.*

*They've frozen them, she said…Any account with an F on it instead of an M. All they need to do is push a few buttons. We're cut off…*

*Women can't hold property any more, she said. It's a new law. Turned on the TV today?*

*No.*

*Night*

*I go back, along the dimmed hall and up the muffled stairs, stealthily to my room. There I sit on the chair, with the lights off, in my red dress, hooked and buttoned. You can think clearly only with your clothes on. It's a funny word, 'job', a man's word. At the time when there were newspapers, millions of women worked. That was before they gunned down the Assembly. They said it was Islamist fundamentalists. I am waiting for the commander whose house I live in. In class, Aunt Lydia showed us a graph where the birth rate drops down below zero. It's not surprising. We learned at the Centre how the atmosphere is saturated with pollution, rays and radiation. There are no more women around on the street where I live, only handmaids like me. The commanders' wives stay in their cars. Sometimes I come across an Econowoman, a poor man's wife. They work as general help, when they can…*

In Atwood's novel, women experience their past as though it were fiction. In order to stay in Gilead, where the war cannot reach, the handmaid submits to her daily role as a sexual slave without rebelling. She complies with the wishes of her commander and his sterile wife, who is confined to private areas. She wanders around as though anaesthetised by the irrational world she lives in, crossing the city centre according to a well-defined custom: an identical route, day after day. That is the third defining feature of Westworld.

But the worst is to come. Violence, conflict, terror and death are an inherent part of the order of Brave Westworld. Isaac Asimov, with his genius, his science, but also with his profound humanism, leads us, slowly but surely, towards this final stage, towards what seems to be complete disorder. However, this titan of science fiction marked himself out from his predecessors, from the literature of the 1930s where robots were conceived along the lines of Frankenstein's destructive creature. He confessed that narrative irritated him. He prefers a world of robots and robotic technology which obeys three laws, the most important of which is: "A robot may not injure a human being or, through inaction, allow a human being to come to harm". However, in the world of The Caves of Steel, humans, who are controlled by extra-terrestrials called Spacers and entirely dependent on robots for goods and services, rebel. In Asimov's world, they have understood nothing. So hate rises in the completely superfluous humans and they turn against the robots, that permanent reminder of their uselessness. Violence takes over, to the author's great sadness, with the humans seeking to regain past glories.

*Everything was set for a mass uprising against the robots. Those masterpieces of innovation had already infiltrated industry, then the services, for years and years. This had meant automatic declassification for the former workers and employees: more and more people from these classes were now subsisting on the strict minimum assigned to sustain life. How could this enormous tide of frustration, from being deprived of earning a living, not turn against the robots, 3D printers and countless connected devices, some with an almost human level of intelligence? This rivalry was intolerable. It was only to be expected that the humans would decide to demolish those soulless upstarts.*

*Baley had heard of similar riots. He had even witnessed one. He had seen robots and devices by the thousand being lifted by a dozen hands, their heavy unresisting bodies carried backward from straining arm to straining arm. Men yanked and twisted at the metal mimicry of men. They used hammers, force knives, needle guns. They finally reduced the miserable objects to shredded metal and wire… the most intricate creation of the human mind, raw material endowed with a super-powerful brain, shaped by deep learning and reinforcement learning. The humans would no longer tolerate the defeat of Lee Sedol by AlphaGo. And sadly they could find only violence as a way of restoring their lost honour.*

Asimov's vision is much more subtle that those texts which portray evil, brutal, destructive robots. It also shows the complex relations between humans and technology; the fear, which moreover is often borne out, that technology could enslave them, indeed annihilate them. But he was never a harsh critic of progress, as he showed in an article written after visiting the New York World's Fair in 1964. In his imagining of the world fifty years later, he barely got anything wrong in his predictions.

Our journey through the greats of science fiction ends here. Perhaps we exaggerated a little and tested the limits: no matter. Obviously, we never set out to criticise humanity's achievements, from the first tools right up to artificial intelligence, or the notion of progress. The vision of "Brave Westworld" is just a mere allegory of the worst outcomes we can imagine of mankind's relationship to technological development. Asimov may be the most hard-hitting on this issue, because he highlights how machines—always presented as a substitute for the drudgery of work—could gradually create a world where humans are no longer in control. Let's not forget that such a story has already happened. At the beginning of the nineteenth century, skilled English textile workers known as Luddites began a violent conflict with manufacturers and smashed up the new machines they suspected of supplanting their work and their income. In the middle of the same century, the Canut silk workers of Croix-Rousse, Lyon, took against the huge machines which had replaced their expertise and had turned them into an unskilled labour force. But we can leave the last word on that to Hannah Arendt, who reminds us that totalitarian regimes transform human beings into superfluous individuals.

## 8.2 Anthropos

Time to move on to a more cheerful version of our future: another exploration of a utopian world, mirroring the first. But this one has the virtue of restoring to politics all of its influence, its power and its ability to see itself in a future built by and for people. We are taking an irreverent approach to classic texts again, but the enduring power and character of these great works mean we can pursue this without playing at being over-the-top futurologists. The world has never been more difficult to predict. In this semi-darkness, the new order that we have named "Anthropos" is the other side of our fantasy future, radically different from the one we have just seen. We will pass through four texts, which also represent four stages, on our travels; they are associated with four values which are also wishes.

The first vital stage is one without which humanity would be condemned to stay in a "state of nature". It is the one philosophers say makes humans human and society society. It is, undisputedly, ethics. So how could we not sustain ourselves with Aristotle's Nicomachean Ethics,[2] even if it means changing the text sometimes to make it fit our purposes better?

*…evidently all lawful acts are in a sense just acts; for the acts laid down by the legislative art are lawful, and each of these, we say, is just. Now the laws in their enactments on all subjects aim at the common advantage either of all or of the best or of those who hold power, or something of the sort; so that in one sense we call those acts just that tend to produce and preserve happiness and its components for the political society.*

*And the law bids us to do both the acts of a brave man* (for example, not to abandon one's free will, not to run away from one's responsibilities)*and those of a temperate man* (for example, not committing the sin of hubris towards oneself and others), *and those of a good-tempered man* (for example, not resorting to violence with words or hands and not to make lying, slanderous or threatening remarks), *and similarly with regard to other virtues and forms of wickedness, commanding some acts and forbidding others; and the rightly-framed law does this rightly and the hastily-conceived one less well. This form of justice, then, is complete virtue, although not without qualification, but in relation to another. And there-fore justice is often thought to be the greatest of virtues, and 'neither evening nor morning star' is so wonderful; and proverbially 'in justice is every virtue comprehended'.*

Justice, from respecting others to shared virtue, is at the heart of the building of a so-called good society, at all times and in any place: that is the very foundation of Anthropos. Unlike Plato, Aristotle promotes an ethics concerned with relationships which does not bypass human vulnerabilities, and which compiles a list of virtues with detailed descriptions. This is a method which could be useful for defining essentially how life should look, this time rooted in a globalised present. Should we now provide a more practical content for this imaginary society?

This brings us back to John Locke. What is a political society if it does not establish common rules, which may, on a global level, be able to devise a functional G20 at last? We are talking about a utopia here which is furthest away from our reality. The issue is actually how to substitute co-operation for conflict. How can we envisage a universe with consensual objectives, without giving back to politics the major role which rightfully belongs to it? It is, in all probability, on this issue that everything depends: specifically, the emergence of a world under construction which does not substitute machines for human will, which does not give the highly complex task of creating compromises, agreements and a common will to technology. And there is no one better than Locke to express it, in the best way:

But because no political society can be, nor subsist, without having in itself the power to preserve the property, and in order thereunto, punish the offences of all those of that society: there, and there only is political society, where every one of the members hath quitted this natural power, resigned it up into the hands of the community in all cases that exclude him not from appealing for protection to the law established by it. And thus all private judgement of every particular member being excluded, the community comes to be umpire, by settled standing rules, indifferent, and the same to all parties; and by men having authority from the community, for the execution of those rules, decides all the differences that may happen between any members of that society concerning any matter of right; and punishes those offences which any member hath committed against the society, with such penalties as the law has established. (Locke 1952)[3]

Locke, unlike his elder Hobbes, has a lot more confidence in modern man and in the social sovereignty that his freedom confers on him. This is how he views the self-establishment of societies, via a voluntary agreement between individuals who are equal and independent. Here, Locke is really talking about re-establishing the dulled power of democratic institutions, which, as we know, people are finding increasingly difficult to adhere to. Daron Acemoglu (2016) denounced the new targets which do no less than form the moral foundations of American democracy. This reconstruction work is urgent if we want normal life to continue, to oppose dangerous, radical ideologies and reinstate political power over those prophets who want to design our future. This is the second rule of Anthropos.

But society does not exclude the individual. Even the name of the Anthropos scenario highlights the need for a new humanism, or of a revitalised humanism, capable of resolving that eternal conflict between the freedom of each individual and the pursuit of the common good. The chosen text is a classic among classics, the very foundation of our social contract, the historic moment when the breach occurred between a world dominated by divine power and the birth of the individual. Rousseau, the philosopher of the contract agreed between people of reason, opens the way towards one version of humanism. Everything must be rebuilt, as we have seen, and this founding text (Rousseau 1762) can help us:

"This sum of forces can be produced only by the combination of many; but the strength and freedom of each man being the chief instruments of his preservation, how can he pledge them without injuring himself, and without neglecting the cares which he owes to himself? This difficulty, applied to my subject, may be expressed in these terms:

"'To find a form of association which many defend and protect with the whole force of the community the person and property of every associate, and by means of which each, coalescing with all, may nevertheless obey only himself and remain as free as before.' Such is the fundamental problem of which the social contract furnishes the solution.

"The clauses of this contact are so determined by the nature of the act that the slightest modification would render them vain and ineffectual; so that, although they have never perhaps been formally enunciated, they are everywhere the same, everywhere tacitly admitted and recognised, until, the social pact being violated, each man regains his original rights and recovers his natural liberty, whilst losing the conventional liberty for which he renounced it.

"These clauses, rightly understood, are reducible to one only, viz. the total alienation to the whole community of each associate with all his rights; for, in the first place, since each gives himself up entirely, the conditions are equal for all; […]

"In short, each giving himself to all, gives himself to nobody; and as there is not one associate over whom we do not acquire the same rights which we concede to him over ourselves, we gain the equivalent of all that we lose, and more power to preserve what we have."

Rousseau has been interpreted in different ways. But the idea persists that politics is based on the existence of a social contract freely consented to by free and responsible individuals. In fact, it constitutes the absolute repudiation of the prophets of modern times. This is the third principle of Anthropos.

Finally, having begun with ethics, we complete our Anthropos paradigm with peace, as a mirror image of the way our Brave Westworld scenario began with institutionalised lying and ended with war. But what is the value of the return of politics and of the contract between individuals, if we do not distance ourselves decisively from violence and therefore from war? Utopia, utopia! Kant takes the lead here (Kant 1795). There is no need for pastiche or additions. He makes the greatest impression by impressing on us his universal story of perpetual peace:

"Peoples or nations regarded as States, may be judged like individual men. Now men living in a state of Nature independent of external laws, by their very contiguity to each other, give occasion to mutual injury or lesion. Every people, for the sake of its own security, thus may and ought to demand from any other, that it shall enter along with it into a constitution, similar to the Civil Constitution, in which the Right of each shall be secured. This would give rise to an INTERNATIONAL FEDERATION OF THE PEOPLES. This, however, would not have to take the form of a State made of these Nations. For this would involve a contraction, since every state, properly so called, contains the relation of a Superior as the lawgiver to an Inferior as the people subject to their laws. Many nations, however, in one State, would constitute only one nation, which is contradictory to the principle assumed, as we are here considering the Right of Nations in relation to each other, in so far as they constitute different states and are not to be fused into one."

Further on, the philosopher adds: "And yet Reason on the throne of the highest moral law-giving power, absolutely condemns War as a mode of Right, and, on the contrary, makes the state of Peace an immediate duty. But the state of Peace cannot be founded or secured without a compact of the Nations with each other. Hence there must be a compact of a special kind which may be called a PACIFIC FEDERATION (foedus pacificum) and which would be distinguished from a mere treaty or Compact of Peace (pactum pacis) in that the latter merely puts an end to one war whereas the former would seek to put an end to all wars for ever. This Federation will not aim at the acquisition of any of the political powers of a State, but merely at the preservation and guarantee for itself, and likewise for the other confederated states, of the liberty that is proper to a state; and this would not require these states to subject themselves for this purpose – as is the case with men in the state of nature – to public laws and to coercion under them".

Kant wrote this famous text six months after the long war between Prussia and revolutionary France. He saw the revolution as an advancement in human history. And yet the Kantian doctrine refuses violent resistance, but is about politics. And as far as Anthropos is concerned, politics, according to him, must "bend the knee before the law". Before him, the Abbé de Saint-Pierre published *Project pour rendre la paix perpétuelle en Europe* (Saint-Pierre 1713) a text which is most often mocked for belonging to the realms of Utopia. For Kant, thought is a matter of law. If he shows himself to be an intransigent pacifist, it is because reason demands the banishment of all war. One may search in vain for a trace of naïve optimism in his statements. There is no moral or theological challenge. If there is a case made, it is incompatible with violence. And only the law, extended to all relations between people and between states, is able to guarantee a peace which is no longer temporary. Although Rousseau, Voltaire and Leibnitz could show themselves to be sceptics and even if history itself offers cruel denials, we must take this laudable risk with Kant.

Kant untangles all the issues here which cause anger: flows of migration, conflicts of religion, borders and influence and the fear that some may dominate others, which today relates to the masters of technology and society. He concludes our construction of an Anthropos scenario: our utopian, illustrated concept of the paths available to us today.

How can these unrealistic caricatures, these imaginary scenarios, be useful to us? The technology prophets are not actually thinking of imposing the terrifying vision described in the Brave Westworld scenario on us. As for politics, it is in such a weak state that it is really incapable of conceiving of the new, twenty-first-century humanism. If we had to weigh up the power relationship between those two parties today, we would be more likely to say that a credible version of the first scenario is the most plausible. The power of the large technological companies is supplanting that of politics: and that is happening today. But the worst outcome is never certain. After the painful transition period we are experiencing, we must hope that in a few years our societies may have the tools to choose to approach the Anthropos scenario. Building on the stylistic exercises above, that is what we will attempt to portray in the final chapter.

## Notes

1. Director of Hatcheries and Conditioning.

2. Aristotle (trans David Ross, revised by Lesley Brown, 2009 ed.) The Nicomachean Ethics, Book V. Oxford World Classics, Oxford.

# References

Acemoglu, D. (2016, November). *American Democracy Is Dying, and This Election Isn't Enough to Fix It. Foreign Policy.* Available via https://foreignpolicy.com/2016/11/07/american-democracy-is-dying-and-this-election-wont-fix-it/.

Asimov, A. (1954). *The Caves of Steel* (1991 ed.). New York: Bantam.

Atwood, M. (1985). *The Handmaid's Tale* (2017 ed.). London: Vintage.

Castel de Saint-Pierre, C. (1713). *Project pour render la paix perpétuelle en Europe* [The Project for Perpetual Peace in Europe] (2014 ed.). Éditions du Linteau.

Huxley, A. (1932). *Brave New World* (2007 ed.). London: Vintage.

Kant, E. (1795). *Perpetual Peace: A Philosophical Essay* (W. Hastie, Trans.) (2015 ed.). CreateSpace Independent Publishing Platform.

Locke, J. (Ed.). (1952). *Two Treatises of Government.* Google ebook. Available via https://books.google.co.uk/books/about/Two_Treatises_of_Government.html?id=K5UIAAAAQAAJ&printsec=frontcover&source=kp_read_button&redir_esc=y#v=onepage&q=political&f=false.

Orwell, G. (1949). *Nineteen Eighty-Four* (2004 ed.). London: Penguin Classics.

Rousseau, J. J. (1762). *The Social Contract* (1998 ed). London: Wordsworth Editions.

# 9

# Re-humanising the World

What an ambitious exercise this was: to celebrate everything that science offers us today to improve the human condition (our health, nutrition, education, etc.) but also to condemn any scientific technology which has lost sight of its creative purpose and its respect for the individual. We have engaged with that task. At present, everything is still to play for, provided the future is not dictated to us by a few technology gurus—our so-called prophets—and that collective political thinking regains the upper hand. Unfortunately, that is not the path we are following today. Not that our politicians are incapable of doing so, but simply because the world is so difficult to understand, the future so uncertain, that they feel paralysed. So we have a simple objective: to set some rules which, if imposed, will give those in power the makings of a strategy.

We will set out five points here which explain the extremely strict parameters governing what our future rules must be. Firstly, they concern the dismantling of those technology monopolies who think they can develop as they wish forever, and therefore somehow dictate the lives of future generations. It will not be the first time such changes have occurred. And we should remember that it was the United States,

© The Author(s) 2019
J.-H. Lorenzi and M. Berrebi, *Progress or Freedom*,
https://doi.org/10.1007/978-3-030-19594-6_9

as the dominant power for a century and a half, which managed to break monopolistic structures which were gradually consigning healthy competition to the dustbin of history. Ethics is another area which does not entail half-hearted decisions. We have access to exceptional amounts of knowledge today, but it is absolutely imperative that we have a well thought out and accepted limit to what humanity can do or undo in the DNA chain. The same goes for privacy. We must be absolutely intransigent on the issue of protecting the individual. Today, he or she is surrounded by sensors on all sides, gathering data on his or her private life, movements and preferences of all kinds. Never in human history has the individual been so condemned to technological imprisonment in this way. As for the final two limits, they touch more on the trajectory of the global economy, as well as global governance. We are in a phase which occurs in all industrial revolutions, where technology destroys more value and jobs than it creates. That is normal: we must simply understand that the transition phase will be long and that we are just at the very beginning of the process. We must therefore invent a type of economic growth which allows us to reverse current trends and to speed up the return to prosperity. The same applies to the role of political power. It is marginalised and sometimes vilified today, such that we cannot imagine satisfying developments resulting from the current state of affairs. Can we contemplate the world functioning without a compass, that is, without democracy? Our real ambition is to give political and intellectual tools to the community, in order to define our future collectively.

## 9.1    Breaking the Technological Monopolies

Above all, let's not imagine that we are in a new situation. Since the beginning of the twentieth century, competition was always conceived, constructed, regulated and defined by the United States. This is because competition was always their first rule of what capitalism of its day should be. It began as it always does, with a powerful reaction, an almost revolutionary movement of little against large, consumers against producers: from a vague feeling that certain people were imposing their law on the great majority. This was the story of pools and trusts in the United States in the second half of the nineteenth century. Everyone knew the major role that John Rockefeller played in the dominant role of the trusts, via Standard Oil and the State of Ohio. It was widely imitated in sugar refining, explosives, cotton oil and so on. All of this led effectively to the disappearance of thousands of small entrepreneurs and operators, and to public opinion reacting against this systematic carving up of the American economy. As we know, it all blew open with the Sherman Act, and then the Clayton Act, which themselves had a turbulent existence, because American economic policy always adapts its principles to the demands of the economic situation. We must bring about the same kind of disruption today, still based on the desire to be respected as consumers, as citizens and as individuals. It should then be supported by strong voices until the apparently immutable order, and dynamic is changed. Who does not remember Harold Green: a minor judge in the Court of the District of Columbia, who would shake the world and force Charlie Brown, the boss of AT&T, to negotiate the break-up of his monopoly? AT&T was the telecommunications empire of the time, a company which was admired by all telecommunications operators around the world and had a million salaried employees. AT&T did everything, from running the network to manufacturing the equipment. AT&T had, through its power and its major role in the American economy, become a sort of icon. And most importantly, most clients, as with a company like Google today, considered the service they received beyond compare. And yet, Judge Green would cause that monopoly to break apart. After an eight year investigation, it agreed to end its telecommunications monopoly. And eighteen months later, through the incredible impetus of this minor judge, American telecommunications were completely turned upside down. Seven local operators, called "Baby Bells", an AT&T long-distance operator and a manufacturer of equipment called Lucent, were substituted for AT&T. In reality, between a customer on the East Coast and another on the West Coast, all communications had been in the hands of the same company and the decision was taken, against public opinion, to cut its activities. The word "cut" appears brutal, but is an apt description of the effects of the competition authorities' decisions. We will use it, because we consider it justified in order to avoid inextricable outcomes in ten or fifteen years' time which would, above all, suit the monopolies which the world could impose on us any day now.

So let's go back to our digital world: the one we talk about so much, which occupies the territory of the media today and offers us exceptional services in our professional and personal lives. But this is not about those aspects of it. We are going to attempt to curtail the abuses perpetrated by those digital monopolies. At least two reasons can be put forward for this approach. The first is to say that these companies have achieved such power—financial, technological, political—that they now have the capacity to make their vision of the world prevail and to impose their choices on governments. These companies now compete with public authorities in their prerogatives, particularly in the areas of health and education. The second reason, a major one, is to prevent too great a concentration of information about us, our lives and our communities in the hands of a single organisation. Why dismantle it, when we could apply the rules issued by the competition authorities? Simply because we are dealing with a very specific resource: data, which is very different from steel or corn. The method we choose for limiting activities, either by separating them or by abandoning some of them, (in a word, dismantling them), necessarily has repercussions on the procedures it entails. Thus we will not use the same methods and dividing lines to dismantle those stakeholders who have become too powerful, as for those who must be prevented from collecting, producing and distributing too much data.

If our ambition is to limit the power of these too-powerful technological stakeholders, then the task is onerous, but possible. This is because it will be necessary to define limits for the activity of all the technological groups who have real financial power at their disposal: enough to buy any kind of business, attract the best talent and influence political power... In such cases, we limit them to their original activities, following the example of the tribunal which ordered the dismantling of Standard Oil into 34 independent companies in 1911.

If we are talking about preventing the concentration of data, we will use other methods of sharing and division. Thus we could order the legal separation of activities, and therefore prohibit the flow of data between different entities. Google, for example, is able to follow the current status of its users across most of its services: Google Search, YouTube, Gmail, etc. and therefore to gather a very large volume of information. This information thus enables them to target advertising at advertisers on Google AdSense. So it could be a case of simply imposing a legal separation between the search engine, the advertising business, Google Maps, Gmail, YouTube, etc. Once these entities are legally separated, the time would be right to implement data transfer regulations rigorously. If a business wants to sell or transfer personal data, it must inform the people concerned by letting them know that their data will be used and what its destination is. Then there are two hypothetical outcomes: where sensitive data is concerned, such as religious or political opinions, or when automated data-gathering is used, acquiring consent would be obligatory. In all other cases, people have a simple right to object. These rules would apply to all stakeholders, regardless of their size.

Very clearly, it is the first reason (limiting excessive power) which takes us away from a simple application of competition laws. We can actually use Senator John Sherman's statement, at the time when he justified his law against Standard Oil, as our blanket rule: "If we will not endure a king as a political power, we should not endure a king over the production, transportation, and sale of any of the necessaries of life".[1] We must expect to see new Shermans in the next few years, ready to stand up for regaining some control over those technological stakeholders.

Let's say it again: the case could readily be argued for more traditional forms of regulation such as control of algorithms, giving competitors access to certain data, making consent by users properly effective and so on. But the history of the world is clearly not made by just applying a more effective set of rules, specifically by putting forward healthy competition. Power is of another nature altogether. We are convinced that faced with this growing power, the demarcation of areas of activity is the only way to restore balance.

Let's take the case of the digital sector, because that is what concerns us first today. Without any exaggeration, we can say it comprises ten professions. The first is managing collaborative networks; the second, search engine activity; and the third is online sales. The fourth is the distribution of content; the fifth the management of self-production. The sixth is the production of digital objects while the seventh is operating systems design. The eighth is database management; the ninth is the building of platforms for disintermediated relationships; and the tenth and final one is the building of widespread communication networks. Everyone will have recognised Facebook, Google, Amazon, Netflix, YouTube, Apple, Microsoft, Oracle, Airbnb and Twitter from this list. What is the common factor in this collection, management, manipulation, processing and marketing of information in different formats? If we examine the core activity of these innovative companies, they all depend on the same procedure, the same backbone. First of all, there is an algorithm which is a little more powerful than the competition, and a slightly more efficient process. Thanks to the network effect, the start-up is propelled into the ranks of must-have apps, thus monopolising a majority of users. The second link in the chain is then to store and accumulate all the data, searches and input from these network users. Whether it involves structured or unstructured data, the company will centralise it, keep it and retain it in so-called data centres, which act just like oil wells. The next stage consists of extracting and refining the data—processing the information—to transform the source into decision-making aids and to propose predictive models. There is a twofold objective, at the very least. On one hand, the company strengthens the quality and relevance of its algorithm a bit more, by offering a more sophisticated and personalised service for its users. At the same time, it can monetise its development costs by: attracting advertisers, offering targeted marketing and ensuring it keeps its leading place in the market by acquiring potential competitors at high prices. The chain is after all quite easy to read and simple in its principle: algorithmic power to attract users; management and conservation of data in order to exploit the data; and predictive models to optimise the targeted marketing algorithms to ensure its longevity. And this original value chain is present in all types of applications, whether it is an Internet search engine, a social network, a video sharing platform, apartment rentals, car sharing, etc.

Every network always starts out with an initial dash for the most volume. This will certainly be brought about by algorithms which are constantly being upgraded, but that is not the core of development. All the large technology firms focus on perfecting their services and widening the nature of the data collected. And adding new kinds of activities to the range of services on offer is enough to achieve that. Monitor your health in real time, manage your physical condition, shop online, use GPS, exchange messages, film and share in real time, etc.… By adding new services which at first glance have nothing to do with the original app proposal, a network can increase its traffic and therefore its monopoly. We are therefore witnessing increasingly vertical integration: a video platform also becomes a content producer, a drivers' platform anticipates having a fleet of driverless cars, a social network envisages virtual reality as a means of communication, a search engine works at increasing life expectancy and an online shopping platform works on delivering its parcels by drones. And at the heart of all those predictive models, there is the private data which users effectively agree to exchange for the chance to enjoy these apps.

The arguments often advanced to prove that it is not necessary to make the authorities intervene to re-establish a consistent competitive market are, on one hand, that the digital markets are uncertain and, on the other, to point out that the markets are unstable. A disruptive start-up with a revolutionary algorithm could emerge at any time, and therefore every technological innovation threatens a company's dominance in a given sector. That is credible today. But tomorrow those ten companies, perhaps fewer, will dominate the world the way AT&T dominated telecommunications. We must not hesitate to prevent this, at any price. As with AT&T, everyone appears to be delighted, everyone sees the immediate benefit of a very efficient communications system, never imagining the configuration of the world economy that is contained in its very principles. So, we are going to propose four changes which will break the development strategy of these companies. Let's be clear, we are talking about various principles which set out one goal, that of limiting the power of companies which are almost monopolies. For example, we believe it is necessary to prevent search engines extending their role by preventing them from accumulating different types of content, in order to avoid a dangerous concentration of information in the hands of one company. In other words, these large technology companies should only be permitted to offer one type of service. It will also be necessary to split the *chaebols* and other conglomerates that accumulate incompatible activities, for example Samsung Electronics and Samsung Life Insurance. It is desirable to limit each company to only using data collected for its own use and to prohibit them from turning it into a saleable asset. The final example would be to order the separation of the production of operating systems/software and devices such as telephones, computers and tablets.

These are just a few of the approaches which demonstrate quite well what we must do in the coming years. All of this takes time and needs resolve, prudence, dialogue and decisiveness. The objective is clear, even if its application is tricky, and as ever, the devil is in the detail. In this initial approach, we are obviously dealing with the large digital companies. Who exactly are we going to talk about? Facebook, Google, Netflix, Airbnb, Twitter and Tencent, who base everything on algorithms and data; Amazon, which relies on logistics and data; Apple and Samsung, who rely equally on design and the production line. Some have diversified, others have not. Some are integrated vertically, others horizontally. Some of them have platform strategies and intervene between existing companies and their customers to reorganise value chains, and others do not.

Let's take the case of Google. Google Search dominates in a large number of countries, and Android, its operating system, is installed in 75% of mobile devices. Google cross-references the personal data collected across its services—Google Search, Google+, Google Maps, Gmail, etc.—which allows it to target advertising, making Google the leading online advertiser. In this way, Google develops a vertical integration strategy: publisher of operating systems (Android and Chrome) intermediary (Google Search, Google News, etc.) content platform (YouTube) and organisational software (e.g. Google Maps). Although Google has given up on the idea of becoming an equipment manufacturer, the company is nevertheless constantly expanding its scope in related areas, for example Google Car, or in activities that are very far from its core business. So it is not surprising that there is a recurring debate over the dismantling of Google. This has not escaped the notice of the European Union, which has been investigating Google since

2010. In 2013, it judged that Google favoured its own content in its search engine. And on 15 April 2015, it officially accused it of abusing its dominant position: a process which also included Android (EU 2015).[2] In June 2017, the EU fined Google €2.4 billion for abusing its market dominance (EU 2017); and in 2018, it fined Google €4.34 billion over Android antitrust violations (Rankin 2018).

In 2014, the European Parliament adopted, by a very large majority, a symbolic resolution which proposed dismantling Google, to separate its almost monopolistic activity as a search engine from its other services (EU 2014).[3] In 2015, Google partly anticipated the debate over dismantling when it restructured the group, creating a holding company, Alphabet, of which Google is a subsidiary. The group's activities which do not directly concern the Internet or mobile communications have left Google and have become fully fledged subsidiaries of Alphabet: Nest, for home automation and connected devices; Calico, the biotechnology company; Verily, another biotechnology company; X, the long-term, multi-project laboratory; Google Capital; GV, the investment fund designed to invest in start-ups; Google Fiber, a project to build a subsidiary network using fibre optics, Sidewalk Labs, dedicated to urbanisation and technology and Jigsaw, a technology incubator aimed at combatting different types of violence and corruption (extremism, terrorism, digital/cyber-attacks, money laundering, organised crime, etc.)

Is it necessary to dismantle an organisation like Google, and if so, how do you do it? Taking into account the separation of activities which has already been carried out through Alphabet, one possibility could be to dissociate Google Search, the search engine, from all the associated activities: Gmail and Google Apps; Google Ads, its advertising business; Google Chrome, the browser; Android, the mobile operating system; Chrome OS, the operating system; YouTube; Google Maps and Google Drive, the information cloud. That's fine, but it's not a match for the problems of next few decades.

We need to think about Facebook in the same, head-on way. There has been a clear issue of improper marketing of data collected from users. Facebook is only dominant in its core business, the social network, with two additions: Instagram for videos and Messenger for instant messaging. Facebook is expanding its capacity for capturing data via Facebook Connect, which is attempting to become an authentication system for a great number of services, and Facebook Payments, which could become a source of revenue through levies on transactions. Today, 95% of Facebook revenue comes from advertising, which is increasingly more targeted. European Information Technology regulation bodies, including the French CNIL (*Commission nationale de l'informatique et des libertés*, or National Commission for Information Technology and Liberties), obtained a commitment from Facebook that it would not merge the data of Facebook users with those of WhatsApp, which Facebook acquired in 2014 for 19 billion dollars. One way of breaking up Facebook could be to dissociate the social network from its extensions in the advertising business.

As for Amazon, the company is only dominant in two markets: online commerce and the cloud, with Amazon Services. Amazon, with Amazon Video, is also in head-on competition with Netflix over video content. In 2015, the European Commission opened a formal procedure of enquiry concerning the distribution agreements for Amazon's digital books (EU 2015). There is a similar company to Amazon in China: Alibaba Group. That group specialises in payment platforms and retail sales, stocking and cloud computing services. The company has also developed its own operating system. Dividing up Amazon and Alibaba could mean separating the online businesses marketing physical goods; the online marketing of virtual goods (videos); services linked to the cloud and the sale of computer equipment (tablets and other devices).

The Microsoft company has escaped being broken up in the United States. In June 2000, Judge Thomas Penfield Jackson ordered the dismantling of the computing giant into two distinct companies: one focussed on Windows operating systems, the other on application software, with no possibility of collaboration between the two companies. For Judge Jackson, "there is no fundamental difference between Bill Gates and John D. Rockefeller, between Microsoft and Standard Oil". But in September 2001, the Department of Justice abandoned the idea of breaking up the group.

Several years later in 2007, Microsoft was fined by Brussels. The European Commission imposed a fine of €497 million, the first in a series of other fines, including one for €899 million. At that time, Microsoft's global share in the operating systems market was at 95%. The Commission demanded that Microsoft "de-group", or separate Windows and Explorer, and obliged it to divulge information about its operating system to its competitors. Since then, Microsoft remains dominant in the PC market with Windows. Nevertheless, Microsoft missed the Internet boat (its browser Explorer has been losing some of its market share since 2004) and the mobile boat; and in the search engine market, Bing is a minor player. Nevertheless, even having lost the leading role that it had before the rise of the Internet, Microsoft remains a giant, which has also taken on a form of vertical integration: operating system (OS); application software; and computer equipment. If it were necessary to restructure the group, it would most definitely be necessary to dissociate the "Windows/application software/cloud" sections from the "Business services" and the "computing equipment sales" sections.

Intel is the greatest global manufacturer of semiconductors, based on turnover. Intel has been prosecuted for abuse of its dominant market position, particularly in Japan in 2005 and in South Korea in 2006, and was eventually fined €1.06 billion by the European Commission in May 2009 (EU 2009).[4] Intel actually undertakes many activities and that is where the real problems lie: computing equipment, with PCs, tablets and drones; data storage with its "data centers", servers, network cards and large-scale storage hardware; computing power with its microprocessors; artificial intelligence and the Internet of Things.

The Samsung Group is one of the leading Korean conglomerates, known as *cheabols*, and holds many subsidiaries, covering business activity in electronics (semiconductors, computing equipment, connected devices, etc.); in engineering and construction; in naval construction and also in insurance, a business which is obviously relevant to digital technology. Here, it would therefore be a case of splitting off the incompatible activities, particularly those which involve Samsung Electronics connected devices and those linked to Samsung Life Insurance.

China's digital giants also focus on several types of activities. Baidu is known for its search engine, which is in an almost dominant position in

China. This company may not have an operating system or web browser like Google, but it is nevertheless developing numerous other services: geolocation, video content, self-driving car technology, cloud data storage, access to financial information, etc. A big digital company, Tencent, specialises in Internet and search engine services, social networks with WeChat and QQ, online advertising, online sales, online games and anti-virus software.

This quick overview of the digital sector in the United States and several other countries is profoundly enlightening. Today, we only notice the incredible dynamism and unbeatable talent. However, it is not about that: it is about our freedom. This résumé is too quick and simplified: it could be criticised on that front. But it has the sole benefit of helping us choose the necessary approach to division, of limits and boundaries, rather than strengthening of regulation, although that is also necessary. This choice is destined to avoid great problems in a few years' time. All of this is obviously very ambitious: but let's be as brave as the American smallholders of the end of the nineteenth century.

## 9.2    Redefining a Global System of Ethics

In the previous chapter, we evoked those illustrious thinkers who set eternal rules for the protection of human society. But we are going to take our inspiration for our recommendations for the future from a new, astonishingly daring genre of fiction: TV series set in a near but totally fictitious future, such as Westworld and Black Mirror. In one random example, The Entire History of You,[5] every individual has a chip implanted behind their ear called "the grain" which enables them to record and play back their memories at will. Liam, a young lawyer, who suspects his wife is being unfaithful, uses and abuses the intrusive technology in order to prove the truth of what started out as just a fanciful suspicion of adultery.

We see the central role of memory, its development and control. This is also the case in Westworld, a series set in an amusement park of that name, populated by androids called "hosts", which rich humans, called "guests" can visit. The guests can act as they please in the park with no risk of reprisal. Little by little the androids—robots who are very close to being human thanks to advanced techniques in anatomical manufacture and artificial intelligence—begin to become conscious of their own existence. Their capacity for memory develops to a point where they remember the violent acts they endure daily: murders, rapes and torture.

What is shocking is that today's reality seems to be gradually catching up with fiction. In 2013, researchers at the Massachusetts Institute of Technology (MIT) succeeded for the first time in manipulating the memories of mice, by creating a false memory from two independent events experienced by the animals. To achieve this, the researchers used optogenetics, a technique devised in the 2000s which enables certain neurones to be activated by exposing them to a light source. In order to make the cells light receptive, they must be genetically modified beforehand to excrete a protein, channelrhodopsin. Mice whose neurones had been modified were placed in an unfamiliar cage, the memorising of which triggered specific groups of neurones. Those cells then excreted channelrhodopsin and became light sensitive. When they were then placed in another cage, the mice's memories from the previous day were artificially reactivated by exposing the neurones to a beam of light while they received an electric shock. When they returned to the first cage, the mice seemed wary and fearful, even though that environment was not linked to any disagreeable event. By reactivating the neurones involved in memorising one cage, while the mice received an electric shock in the other, the researchers artificially associated two distinct events, manufacturing a false memory. According to Pierre-Marie Lledo, "Memory processes would have been preserved intact between species throughout evolution, from worms to human beings" (Gardier 2013).[6] He says: "fictitious memories are just as effective as real memories at triggering emotional responses. In both cases, the same neuronal circuits are involved. The boundary between what is real and what is imaginary is fading away".[7]

All of this is fictional and real at the same time, which enables us to reflect seriously on what the basis should be of a global system of ethics designed for this ongoing revolution. Because this is only one small example of the transformations we can envisage being carried out on humans: brain grafts, all sorts of genetic manipulation, and cloning like Dolly the sheep, the first experiments of which we have seen in China. We know that CRISPR-Cas9 was tried for the first time on a human being in China in October 2016. It involved editing the human genome and injecting cells containing modified genes into a patient. The trial has a good objective, to beat cancer. We can see that there is a desperate need for new ethical rules.

What does Laurent Alexandre, one of France's foremost ambassadors for transhumanism, have to say about this? According to him, humans will acquire unlimited abilities—"powers", even—thanks to the NBICs: nanotechnology, biotechnology, information technology and cognitive science. They will be able to modify their genomes and their nature, turning themselves into "man-gods". He says the man who will live for 1000 years has already been born, and that in twenty years we can expect some people to have intra-cranial implants installed, connected to the Internet.

Obviously, these statements are really much too radical, but they also help clarify the present difficulties and those to come. That is why we cannot conceive of anything but a radically new development, consisting of ethical rules imposed worldwide. This might seem totally unrealistic, but in fact, it is not. Two examples demonstrate that we can draw up treaties at a global level which impose very strict rules on the entire planet. First of all, let's remind ourselves of the International Atomic Energy Agency, or IAEA. Two nuclear tragedies occurred at the end of the Second World War; in 1953, the American president, Dwight D. Eisenhower, proposed the creation of an international agency with a very precise objective: to control the use of nuclear material. Genetic engineering presents us with a similar challenge. And that is actually what makes us ambitious and optimistic. We saw once again recently to what extent the IAEA could impose its will. The agency was properly activated in 1957, and since then, it has appeared as the absolute guarantor of ethical rules at a global level. And for another issue, there was another method: a treaty which was also constituted at a global level. We are talking about the Antarctic Treaty, which in 1959 set definitive rules of ownership of the southern continent, in order to ensure that this region of the world would continue to be used exclusively for peaceful purposes. "Antarctica shall be used for peaceful purposes only. There shall be prohibited, inter alia, any measures of a military nature, such as the establishment of military bases and fortifications, the carrying out of military manoeuvres, as well as the testing of any type of weapons".[8] And here too, the treaty is implemented and is adhered to by all on a decisive basis. So let's return now to what form a genetic engineering treaty and agency could take.

Considering the serious consequences of such research for humanity, any convention or treaty should institutionalise, at an international level, the good practices defined below. Correlating to and taking inspiration from what we name and pursue as crimes against humanity, the definition of a crime or a violation against the human species should be the subject of a consensus, in order to make the stakeholders concerned responsible and to condemn those who violate such a convention.

From there, we can already envisage several proposals. The first group of recommendations would concern activities in the field of biology. First of all, an international moratorium prohibiting genetic manipulation of human embryos should be implemented. Moreover, various obligations should be put into effect in order to better regulate research activity. Laboratories and their personnel carrying out genetic manipulation, that is gene therapies, and research on animal models, should be obliged to declare this to the public authorities. It should be compulsory to obtain permission in advance for trials and research using molecular scissors, CRISPR-Cas9, or genetic transcription methods; it should also be compulsory for biological products resulting from research to be traceable with associated documents and security measures and protection against biological hazards. In order to ensure the effective implementation of these regulations, the creation of an international agency, modelled on the IAEA for atomic energy, clearly appears to be desirable. A second set of recommendations would comprise strict prohibitions on any activity that would lead to a living/technology hybrid or to altering human consciousness, such as artificial procedures to increase sensory or cognitive capacity, falsifying memories, transferring consciousness to an artificial medium and prenatal modification of genes without medical need.

We can see that the field of intervention of this agency will be very significant and the debates to come will be very animated, if not violent. But the quantity of problems that are posed today (and the even greater number tomorrow) in this highly sensitive field of genetic engineering is such that we cannot wait. We urgently need a forum for international reflection and subsequent decision-making in this field of genetic engineering.

## 9.3   Restoring Real Privacy

This may be the area which touches us the most at present and on a global scale these days. We are all aware of being constantly followed, watched and influenced by companies who, by supplying us with services, determine our profiles for us and use them to optimise their sales and marketing campaigns. But this is just the very beginning. And, unsurprisingly, all of this is done without our consent. Geolocation is no longer intermittent, but permanent. But the area where a terrifying step has been taken is in the control of our own wishes. Knowledge about the individual is and will be on such a scale that it will end up defining the individual's own behaviour. We can actually talk about a self-fulfilling mechanism: the near-final loss of our freedom of choice as individuals.

In reality, there are two distinct problems: one at the level of the behaviour of states and the temptations they can experience to implement different forms of surveillance, and the other of digital businesses, for which it is no longer a case of temptation, but reality. At the first level, that of states, our perspective is there could be global conventions, similar, for example, to the one which prohibits chemical weapons, that is state aggression towards the physical integrity of individuals. In our case, the issue is moral integrity. Adapting that concern in order to protect the individual or the community from state intrusion would involve prohibiting states from developing, producing, finalizing, acquiring, storing and holding these intrusive information systems. Each state is therefore obliged to put in place national legislation which extends the provisions and the prohibitions of the convention to individuals and companies.

But let's go back to the much trickier problem of the dominant technological companies gaining increasing control over the individual. In the future, geolocation will be so widespread that, on the one hand, to withdraw from it would appear suspect and antisocial, and on the other hand, its permanent nature would make precise and complete information available on the behaviour of each individual. The CNIL gives us another example. According to Emmanuel Kessou, Big Data, which gives us a rating according to the goods and services which we want to acquire, will eventually deprive the individual of their freedom to choose the company they want to do business with and hand the choice to the company instead (CNIL 2012).[9] We can see that the nature of the intrusion is changing, depriving the individual of some of his or her self-determination and giving the technological environment the role of final arbiter. One could even imagine that in a world where the individual can be built or rebuilt partially out of prostheses, deprived of his or her freedom of choice, technology could in effect design a kind of android. This evil vision of human history is not inevitable, and there is no deterministic force controlling this gradual power grab over humans. But what is certain is that our rules are still very naïve. Take the example of the CNIL in France. Its aim is laudable. It consists of setting rules for the absolute separation of different sets of data available about a person, because it is actually the combination of those data sets which curbs freedom. And the CNIL absolutely does what it can. It monitors and sanctions every abuse, so that today it is impossible for a company and/or an insurer to know an individual's health data. So all is not lost. The "Privacy Shield", signed in 2016 between Europe and the United States to protect personal data, is the start of a legal framework. Let's also remember that the protection of personal data is a fundamental right which is guaranteed by Article 8 of the European Union Charter of Fundamental Rights (EU 2016).[10] Nevertheless, there are worrying limits at the moment on the willingness of states to apply this law, in particular to matters regarding the means of data collection and the effectiveness of the right of appeal, while the populations and companies concerned are increasingly worried by information being revealed, or which is susceptible to being revealed, by various channels.

This is why public debate and action are needed in this area. It is usual for a person whose abilities are limited for health reasons to be treated differently by their employer. But we can clearly see that this could inevitably lead to insurers moving towards selection on the basis of risk, perhaps one day going as far as seeking each individual's genetic data. This possibility would be just the first step towards a limitless expansion of the information known about individuals in all areas of life, facilitated by a dominant technology. The data collection companies would be the natural stakeholders in this development, even if they are all conscious of the problems it poses, albeit sometimes in a very naïve way. Remember, the CNIL, instead of heavily sanctioning Google, obliged the platform to publish the judgement the CNIL delivered on it. Effectively, the institution sanctioned Google for having merged the data of all of its subsidiaries and services (YouTube and Gmail, among others) with no legal basis for doing so, while the law on Information Technology and Freedoms had fixed rules on confidentiality. The sanction appeared to have very little deterrent effect.

But as we have seen, there is much more to privacy than that. And if we do not call a permanent halt to all forms of intrusion, whether digital or genetic, Asimov's nightmare will come true. In order to outline an anti-intrusion policy, we can first go back to four major principles suggested by the French Council of State in 2014. The first is the right to self-determination regarding information technology. Issued by the German Federal Constitutional Court, this law aims to "guarantee in principle the capacity of the individual to decide on the communication and use of his or her personal data". Next, the principle of the neutrality of the operators of electronic communications must go hand in hand with an obligation of loyalty for platforms. The platforms must act in good faith in the classification and referencing of the data, making no alterations and not diverting it towards uses unknown to the user. The third principle says it is necessary to implement the right to delisting (the right to be forgotten) confirmed by the European Court of Justice in 2014. Finally, it is important to cover identification numbers with sufficient guarantees of anonymity and to categorise the retention of data in all files, for example according to the seriousness of the crime in police files. But make no mistake: implementing all of this will be very difficult.

Why? Because, the economic model of the likes of Google, Yahoo and Facebook consists exactly of identifying users and knowing everything about their consumption habits in order to sell on their data. In exchange for which users have access to a service that not only functions perfectly, but what is more, is free. That is how these companies function. And this business model, which consists of knowing everything about your users, is exactly the same as that used by intelligence services, except that in the case of the technology firms, the motivation is obviously commercial. There is a real connection between the public intelligence organisations, such as the NSA, and the web giants. Clearly, when a bridge of that nature is built between two systems, intrusion into citizens' private lives becomes very probable.

However, the years 2015 and 2016 saw a significant change appear throughout Europe. The fundamental principle which says the role of public authorities is first of all to ensure global security, so that the population can enjoy its rights, rather abruptly gave way to the idea that the public authorities can restrict our rights in order to ensure our security. The consequence of this change, which stemmed from the risk of attacks, was an insidious but distinct redefinition of the boundary between the powers of the state and the rights of individuals. Some members of the European Union reacted to terrorist attacks by proposing, adopting and implementing anti-terrorism measures which strengthened executive powers and exposed the entire population to heightened surveillance by public authorities. This was all understandable and legitimate, but it cannot serve as a long-term model.

In a way, this new situation has motivated various people to reflect on what has enabled the individual to protect themselves from different forms of intrusion. Independent experts from the technology sector have listed the most urgent decisions to be taken from this perspective: reduce the number of new opportunities open to service providers for processing electronic communications metadata; make it compulsory to process the data anonymously whenever possible; by default, block third-party cookies and all other third-party access to the device;

prohibit the tracking of devices in certain physical locations, whether they be private, like a shop, or public, like a park; expressly prohibit highly intrusive tracking practices; time-limit the exemptions allowed to states for the electronic surveillance of individuals, even if it is done for reasons of national security; treat the violation of the protection of final equipment, the hacking of a device or the tracking of that device in an illegal manner as a violation of the law. If all of these measures were applied, it would represent real progress.

But it seems to us that we must go much further, in relation to companies as much as to private individuals. Specifically, we can conceive of a separation between the activity of data collection and the use/resale of the data for commercial ends and the prevention of the whole value chain of a given activity being controlled by one digital stakeholder, in order to avoid abuses of dominant market position. As we mentioned, dividing up the digital giants—AT&T style—would allow us to keep better control over the consequences that each instance of data manipulation brings. Finally, in order to avoid that dangerous liaison between public and private life, access by administrations to data collected by technology firms would be prohibited without specific authorisation.

Are these objectives enough? Definitely not, but freedom is something we must fight for; it is up to individuals to regain for themselves the taste for and the reality of their individual freedom.

## 9.4    Putting Technology Back in the Service of Prosperity

We know that today, we have been plunged into a technological movement which is only in its early stages, which is clearly having negative effects on employment at the moment. Today, it feels as though digital technology is at the heart of this development. But in reality, it is not. What we see is just the distribution and marketing channels undergoing a systematic process of disintermediation, which changes patterns of consumption somewhat. But we are still a long way from a radical transformation of the economic world. We can envisage—and that is what is important to us in this work—that the development of science in the areas of space, of the infinitely big and the infinitely small and of molecular biology, will really turn our modes of production and consumption upside down. And digital technology will be involved in it all, as will biology and astrophysics.

This is how technical progress works. At the core of economic thinking for the last two centuries, it tirelessly shows its two faces: on one side is progress, growth, the ending of hard labour, the invention of science and of life-changing products. On the other is the image of the disruption of the markets, instability, unemployment and the enslavement of the individual to goods they do not want. We are taking a resolutely optimistic approach, because industrial revolutions have helped societies to advance in worlds where intelligence is valued more and more. Any examples of excess and suffering were only caused by the lack of regulation of the effects of these radical developments. Nevertheless, we have described the emergence of systems, observed in the past, which are leveraged by new techniques, by newly discovered ways of organising labour, by new forms of consumption. These are called "techno-industrial systems" (Bourlès and Lorenzi 1999) aligned here with the thoughts of Schumpeter and others who have never considered technological development as a simple continuity, but rather as having certain favoured periods, concentrations of new techniques and of accidents. Put simply, we must add some other changes to this vision shared by a number of economists: those which affect the nature of work and consumption.

The specific feature of today and tomorrow is the return, despite appearances of modernity, of a nineteenth-century society. We have forgotten too quickly that the world has, for most of its existence, experienced periods of slavery and servitude. Today and tomorrow, access to knowledge will separate us into lords or serfs. It is strange to use this type of language, but how better to describe the coming years, if not by observing the polarisation of our labour markets between very high qualifications on one side and very low qualifications on the other. No one is disturbed by it, since our prophets have described a society to us that only Marx could have conceived, "to each according to his needs". Here, too, the worst outcome is never certain, and the trap which a huge number of men and women find themselves, without hope, without a future, for them and their children, is by no means inevitable. The real danger is that technology should appear to be an objective basis for discriminating between human beings. And that deep down, no policy could oppose that twisted point of view. Science seems unstoppable because it now plays the role of Supreme Being. Here too, we must stand up to its obvious statements, its soothing arguments, its monstrous inequalities and this rampant dehumanisation! The current responses, in terms of the fight against inequalities, however nice they may be, are not fit for purpose. We will not combat the dynastic reality of our society by adding a layer of taxation or increasing taxes on capital. Knowledge is actually the absolute value of the twenty-first century. That is why we are formulating four propositions, possibly insufficient, but which must at the very least to be implemented on the scale required. A first measure consists of understanding and then adapting to the fact that professional roles must be subject to change in tomorrow's technological world, for every one of us, which will be incomparable to what we know today. Several policies could be implemented in order to meet this new problem head-on. But we believe we must decide on a second chance society, which lets everyone have a period of training during their professional life, enabling the possibility of a change in direction. This only makes sense if that possibility is offered to everyone. It also means that everyone must, in order to find their way during their transition period, have a fairly precise idea of all the jobs which are available to them, and that the period of training will qualify them to go into this new area of work. Finally, this means that we must remove glass ceilings, which would prevent, for example, a nurse from becoming a doctor, whatever additional training was undertaken.

Following that, technology today has allowed the expansion of entrepreneurship, alongside a still significant salaried workforce. The social security system must combine the systems covering risks for both types of worker, in all aspects of social security: health care, retirement, unemployment and critical illness cover.

The third measure depends on digital innovation. We can use technology to systematically observe each person's talents from childhood and through their whole life, and to constantly open up all the most suitable training for them, based on each person's knowledge, aptitude and preferences.

Finally, why not share knowledge in a much wider manner than the choices the large technology companies offer us, based on our preferences, which are supposedly detected and analysed and which enclose us in a predetermined world? We really must rebel against this algorithmic containment, which is nothing other than the technological translation of social inequality. The creation of digital communities imprisons individuals inside what is supposedly reserved for them. We can conceive of randomised systems, which would allow each person to access culture offered as a shared asset, accessible to all.

## 9.5   Towards a New Public Authority

How can we imagine for an instant that this redefinition of the way our society works, these constraints to be imposed on companies, these strong limits on technology's major players, will not fundamentally transform the social and political contract which has connected citizens to each other for almost two centuries? Many people imagined that the Internet was going to create a new form of democracy. There was a feeling that we could establish a sort of electronic agora [ancient Greek marketplace]. That is the position of people like Patrice Flichy, or of Dominique Cardon. Others envisage very different developments. They see in the setting up of these digital communities the risk of new forms of communitarianism which could present many dangers. This is moreover the only area of this emerging worldwide technological industrial revolution where digital technology plays the fundamental role, because it removes borders and social groups. It gives absolute power to whoever manages the flow of information. Now, we have seen that the protection of the individual in all areas of technology can only be managed at a global level if we want the desired political reaction to really be applied. Finally, in a world where the virtual often prevails over what is real, we can easily imagine the individual wishing for a renewed strong connection to his or her physical and political environment, language and the institutions he or she is close to.

What is public authority in this new environment, if not, as ever, the collective will to decide the future one wants, made by a majority of individuals in a clearly defined physical area? The first condition necessary for that to happen is that dialogue must truly be established between a legitimate power and recognised checks and balances. The regrettable weakness of the public authorities in technological matters makes them unfit to oppose the increasing powers of the technology firms, which explains how the narrative of these companies has taken hold so easily. The absolute priority is therefore to give administrations and agencies the ability to master all aspects of present and future developments and to be able to present a credible counter-argument to those who are in control of technology today. But in societal debates, political assemblies are often much freer and confirmed in their positions than states themselves. Assemblies themselves must also have a sufficient level of technological competence, because they will be the major players in the analysis and subsequent control of the sometimes unacceptable applications of technological innovations desired by certain large companies. One example, which is completely marginal but very symbolic of this new role of assemblies, was that of the ability to patent software. The hero of this refusal to take the American position on this topic was, surprise surprise, the European Parliament, which prohibited the American-style legalisation of software products.

But above all, the public authorities must, in the coming years, operate with a very different perspective from the past: they must become an absolute fortress protecting the individual facing the eventual consequences of technology. What level would this be on? Imagining worldwide institutions is always complicated, but we can hope that, on the three points mentioned (the struggle against monopolies, the creation of a system of ethics and the protection of the individual) those would be rules adopted by all and implemented and enforced by the governments of every country.

We are starting from the reality of current and future power relationships. They have and will continue to have a tendency to favour economic stakeholders rather than political ones. It is therefore essential, because of this new situation created for this acceleration in science, to reaffirm what today could appear to be simple nostalgia: the primacy of political institutions in relation to the economic sphere.

Public authorities must also ensure their new legitimacy over a third function. Technological developments have many surprises in store for us and each one of those surprises brings new questions. The tradition has always been to deal with these subjects retrospectively rather than in advance. The case of nuclear technology is a perfect illustration of this. No decision which entails technological choices should be made today without holding a debate beforehand at a high level, genuinely including all stakeholders, based on rigorous and accessible scientific documentation. It falls to the public authorities to impose this constraint, which is desired by all.

Finally, we must have intergovernmental co-operation in all the areas which now concern all of humanity. However, we must avoid excessive use of the desire to co-operate at any price, which can be counterproductive. We must therefore limit the areas to be dealt with co-operatively to those for which there are strong externalities, obviously in the technological and environmental sectors.

Armed with these four principles, the public authorities, which are today marginalised, widely absent from the debate on the future of our society, making way for the prophets who have emerged from technological disruptions at all levels, can regain their legitimacy, which is so important for our lives.

* * * * *

The status of economists is decidedly difficult to maintain today. For years, they were considered as the most dependable interpreters of a global society which was doing well, and everyone listened with confidence to their analyses of the past and their predictions for the future. In a way, they were the gurus of a fast-growing world, where the emerging countries, aiming to catch up fast, took the baton from the prosperous, ageing countries. And then, all of their predictions, including those of accidents like the 2008 crisis, discredited their arguments in some way. This pushed some towards excessive prudence, and others towards highly technical works which were often useless. Basically, there was a place waiting to be filled by those who could point the way ahead and express what is desirable for everyone. Chance decided that this break in the chain of legitimate discourse was accompanied by an exceptional boom in scientific progress and in the very rapid development of extremely effective technologies designed for all of our daily lives. At the same time, the difficulty in taking control of the world economic situation and its transformations was sowing the seeds of doubt in the ability of politicians to act. An incredible situation emerged from all of this, which we have attempted to describe in this book: the takeover of public discourse and power by the leaders of the large technology firms. It is they who, faced with the void left by the economists and politicians, relying in an unwarranted way on the idea that they were the proper voices of scientific progress, considered that they should decide what the coming decades would be like. We are admirers of, among other things, those scientific breakthroughs in all fields: genetics, astrophysics, energy and digital. But like every scientific advance, those applications must be governed by rules adopted by consensus, often on a global scale. And it falls to the politicians, in all their forms, to set those rules which will be so decisive for the coming generations, quite simply because they are, as a general rule, the expression of our collective will. Scientific progress, as was almost always the case throughout human history, must create conditions of economic and social progress conceived for the good of all and must not be the subject of prophecies from anybody. That is why the conclusions of this book are so constraining: simply because the future of society is very complex, consisting of trial and error, research and discussion; and because we have probably not found ourselves confronted with disruption on this scale for centuries.

## Notes

1. Congressional Record, vol. 21 (1890), p. 2457.
2. European Commission "Abuse of dominant position" The Commission sends statement of objections to Google on comparison shopping service; opens separate formal investigation on Android. Available via http://europa.eu/rapid/press-release_IP-15-4780_en.htm

3. European Parliament resolution of 27 November 2014 on supporting consumer rights in the digital single market (2014/2973(RSP)). Available via http://www.europarl.europa.eu/sides/getDoc.do?pubRef=-//EP//TEXT+TA+P8-TA-2014-0071+0+DOC+XML+V0//EN.
4. Antitrust: Commission imposes fine of €1.06 bn on Intel for abuse of dominant position; orders Intel to cease illegal practices. Available via http://europa.eu/rapid/press-release_IP-09-745_en.htm.
5. Zeppotron, Channel 4 (2011) Black Mirror, Episode 3, Season 1.
6. Cited in Gardier S (2013) De faux souvenirs implantés chez les souris (False Memories Implanted in Mice). *Le Figaro*. Available via http://www.lefigaro.fr/sciences/2013/08/11/01008-20130811ARTFIG00139-de-faux-souvenirs-implantes-chez-des-souris.php.
7. Ibid.
8. The Antarctic Treaty, signed December 1, 1959, Article 1.
9. Quoted in: La *"dictature" des algorithms: demain, tous calculés?* (The "dictatorship" of algorithms: will everything be calculated tomorrow?) CNIL, Cahiers IP, no. 1, Vie privée à l'horizon 2020 (Private Life by 2020), 2012, p. 19. Available via https://www.cnil.fr/sites/default/files/typo/document/CNIL-CAHIERS_IPn1.pdf.
10. The EU rules on the protection of data puts citizens in charge—June 1, 2016. Available via http://www.europarl.europa.eu/pdfs/news/expert/background/20160413BKG22980/20160413BKG22980_fr.pdf.

# References

Bourlès, B., & Lorenzi, J. H. (1999). *Le Choc du progrès technique*. Paris: Economica.

European Commission. (2015). Press Release. Available via http://europa.eu/rapid/press-release_IP-15-5166_en.htm.

European Commission. (2017). Press Release. Available via http://europa.eu/rapid/press-release_IP-17-1784_en.htm.

Rankin. (2018, July 18). Google Fined £3.8bn by EU Over Android Antitrust Violations. *The Guardian*. Available via https://www.theguardian.com/business/2018/jul/18/google-faces-record-multibillion-fine-from-eu-over-android.

# Index